CALF REARING

A practical guide

Second Edition

John Moran

Copyright © Natural Resources and Environment, Victoria, 2002

All rights reserved. Except under the conditions described in the *Australian Copyright Act 1968* and subsequent amendments, no part of this publication may be reproduced, stored in a retrieval system or transmitted in any form or by any means, electronic, mechanical, photocopying, recording, duplicating or otherwise, without the prior permission of the copyright owner. Contact LANDLINKS PRESS for all permission requests.

The author asserts their moral rights, including the right to be identified as the author.

National Library of Australia Cataloguing-in-Publication entry

Moran, John, 1945– .
    Calf Rearing: A Practical Guide
    Second edition
    Bibliography
    ISBN 0 643 06766 3
    1. Calves – Australia. I. Moran, John, 1945– Calf Rearing: A Practical Guide. II. Title.

636.2070994

Published 2002, reprinted 2005, 2007, 2009, 2015, 2016, 2019

Published by and available from:
Landlinks Press, an imprint of CSIRO Publishing
Locked Bag 10
Clayton South VIC 3169
Australia

Telephone: +61 3 9545 8400
Email: publishing.sales@csiro.au
Website: www.publish.csiro.au

Designed by James Kelly
Set in 11/14.5 Minion
Typeset by Mike Kuszla, J & M Typesetting Services
Printed in Australia by Ligare
Front cover photograph courtesy of Dairy Research and Development Corporation.
Back cover photographs courtesy of The Land Newspaper.

Disclaimer
While the author, publisher and others responsible for this publication have taken all appropriate care to ensure the accuracy of its contents, no liability is accepted for any loss or damage arising from or incurred as a result of any reliance on the information provided in this publication.

July19_01

Foreword

Calf rearing systems play a critical role in ensuring an ongoing productive and profitable dairy herd. Systems vary from complex, labour intensive approaches to simple ad lib provision of milk and pasture.

Many dairy farmers could raise additional heifers and increase the opportunities for genetic gain in their herds or participate in other developing market opportunities. Other dairy farmers could improve their calf rearing techniques to ensure their replacements reach the necessary size and maturity to be useful first year players.

It is important that dairy farmers understand some of the biology of calf nutrition and development to assist making the decision on what is the most appropriate for each farm.

This book provides farmers with the necessary information in an easy-to-read format to understand calf biology, evaluate alternative feed resources and to develop a full management strategy for calf rearing. The first edition, published in 1993, has now been updated to include the latest information on calf rearing, specifically for Australian conditions.

This information will assist calf rearers to maximise their returns from investment of time and cash in tomorrow's herd.

Peter Owen
President, United Dairy Farmers of Victoria, Melbourne

CONTENTS

Foreword   iii
About the author   x
Acknowledgments   xi
Chemical warning   xiii

Chapter 1 Introduction   1
    The principles of good calf rearing   2
    The high costs of calf rearing   4
    An outline of this book   5
    Texts for further reading   6

Chapter 2 The principles of digestion of feed in calves   7
    The calf digestive tract   7
    The milk-fed calf   9
    Rumen development and the process of weaning   11
    The role of roughage in the weaning process   12

Chapter 3 The importance of colostrum to newborn calves   15
    Changes in recommendations on colostrum feeding   16
    Colostrum quality   17
    Identifying and storing good quality colostrum   19
    Feeding colostrum to newborn calves   20
    How to stomach tube a calf   21
    Commercial aids to calf rearing   22
    Results from overseas research on colostrum feeding   24
    Timeliness of colostrum feeding   26
    Summarising good colostrum feeding management   27
    Value-adding colostrum   29
    Financial benefits from good colostrum feeding practices   29

Chapter 4 Nutrient requirements of calves — 31

- Water — 32
- Energy — 33
- Protein — 35
- Fibre — 38
- Minerals and vitamins — 38

Chapter 5 Obtaining the calves — 41

- Sources of calves for purchase — 41
- Selection of calves — 42
- Price and availability of calves — 44
- Legislation regarding marketing, transport and slaughter of bobby calves — 44
- Other guidelines for transporting bobby calves — 48
- Bobby calf declarations — 49
- On arrival at the rearing unit — 50

Chapter 6 Milk feeding of calves — 52

- Teaching calves to drink — 53
- The choice of liquid feeds — 53
- The choice of feeding methods — 56
- How much milk to feed — 59
- Other aspects of artificial rearing — 60
- Multiple suckling using dairy cows — 64

Chapter 7 Calf milk replacers — 70

- The composition of milk replacers — 70
- Describing quality of milk replacers — 73
- The nutritive value of milk replacers — 74
- The relative cost of milk replacers — 76
- Using milk replacers to rear calves — 77
- Examples of several milk-replacer rearing systems — 79

Chapter 8 Solid feeds for calves — 82

- The nutritive value of solid feeds — 83
- Feed intake and calf performance pre-weaning — 88
- Feed intake and calf performance throughout the rearing period — 89
- Criteria for weaning calves — 91
- Concentrate mixtures for early weaning calves — 92
- Formulating rations for weaned calves — 95
- The role for pasture with weaned calves — 97
- Special requirements for pink veal systems — 99

Chapter 9 Communicating with the calf — 102

- Signals to watch for from the calf — 103
- Changes in normal calf behaviour symptomatic of stress — 104
- Visual changes in calves symptomatic of stress — 108
- Understand how calves and heifers react to people — 113
- Communicate with your calf rearer, too!! — 114
- Contract calf rearing — 115

Chapter 10 Disease prevention in calves — 117

- Calf scours or neonatal diarrhoea — 118
- Pneumonia and other respiratory diseases — 124
- Pulpy kidney and other clostridial diseases — 125
- Internal parasites and their control — 126
- Johne's disease — 127
- Other diseases in calves — 129
- How to recognise sick calves — 133
- Calf management and disease — 134
- What should you do with sick calves? — 136
- Maintaining a healthy calf shed — 137

Chapter 11 Housing of calves — 139

    Types of shelter — 139
    Management considerations — 141
    Physical comfort of calves — 142
    Types of flooring — 143
    Feeding and handling facilities — 145
    Calf scales — 146
    Cleaning and sanitising feeding equipment — 147
    Calf sheds and children — 148
    Summary — 149

Chapter 12 Welfare aspects of calf rearing — 150

    Government codes of acceptable farming practice — 151
    Additional management practices not included in the above code — 155
    Australian Veterinary Association's policy on calf welfare — 156
    Key issues identified by the Animal Welfare Centre — 158
    Public lobby groups — 159

Chapter 13 Post-weaning management — 161

    On-farm rearing of replacement dairy heifers — 161
    Benefits of heavier heifers — 164
    Target live weights for growing heifers — 165
    Feeding heifers to achieve target live weights — 166
    Using dairy stock for beef production — 169

Chapter 14 Economics of calf rearing — 174

    Costing different feeds for calf rearing — 175
    Other costs to consider in calf rearing — 176
    Categorising calf and heifer rearing costs in the US — 178
    The cost of diseases in calves — 179
    A case study of cost savings through changing milk feeding systems — 179
    Comparing different systems to calculate total feed costs for the first 12 weeks of rearing — 180

Chapter 15 Best management practices for rearing dairy replacement heifer   184

    What makes a good calf rearing system?   184
    Monitoring your calf rearing system   187
    What is best management practice and quality assurance?   189
    Checklists for quality assurance when rearing dairy replacement heifers   190

Appendix 1 John Moran's 10 golden rules of calf rearing   *199*

Appendix 2 John Moran's golden rules of heifer rearing   *201*

    Targets   201
    Feeding   202
    Management   202

Appendix 3 Glossary of technical terms   *202*

Appendix 4 Further reading   *209*

    Textbooks and manuals   209
    Useful websites on calf rearing   210
    Electronic discussion groups   211
    Calf rearing newsletter   211

About the author

John Moran is a senior research and advisory scientist at Victoria's Department of Natural Resources and Environment, located at the Kyabram Dairy Centre in northern Victoria. For the last 21 years he has been researching and advising farmers on more intensive dairy and beef systems for southern Australia, including the rearing of dairy calves for meat production. Prior to this, he researched beef production in northern Australia, South-East Asia and England.

John graduated in 1967 with a Rural Science honours degree from New England University at Armidale in NSW, followed by a Masters degree in 1969. In 1976, he obtained a Doctorate of Philosophy in Ruminant Production from the University of London, Wye College in England. His professional career has been devoted to improving the profitability of cattle producers in the fields of nutrition, rearing young stock and more recently in dairy farm management. John has published more than 200 research papers and advisory articles.

John has also written several farmer manuals on dairy and beef cattle nutrition, veal production and maize silage. The first edition of *Calf Rearing: A Guide to Rearing Calves in Australia*, published in 1993, sold more than 10,000 copies. His book, *Forage Conservation: Making Quality Silage and Hay in Australia*, published in 1996, is now a set text for undergraduate study in several Australian universities. His most recent book, *Heifer Rearing: A Guide to Rearing Dairy Replacement Heifers in Australia*, was published in 2001, in collaboration with Douglas McLean, a fellow graduate in Rural Science. The revitalised interest in the management of young dairy stock resulted in a request for him to revise *Calf Rearing*.

John was one of the initiators of the Australian Maize Conferences, a triennial forum for maize growers and users. He also holds executive positions in national agricultural science and animal production organisations. In recent years, John has been invited to develop and present dairy farmer and advisory training programs throughout South-East Asia.

His wide knowledge of Australia's dairy and beef industries stands him in great stead for this the second edition of his book, written specifically on young dairy and beef stock management in Australia.

Acknowledgments

The deregulation of Australia's dairy industry, in June 2000, has created a 'sea change' in the evolution of systems for producing milk in Australia. In states where milk quotas have been removed and milk returns have generally decreased, farmers have had to re-evaluate their cost structures, while in other states previously without such milk quotas, many farmers have expanded their milking herds to increase farm incomes. In recent years, increasing emphasis is given to the feeding and managing of their replacement heifers as well as the milking cows. Farmers are now better informed on the nutrient requirements of their young stock, the importance of proper facilities for calf rearing and also the many diseases that can affect their heifer calves. Producers often question the wastage through the wholesale slaughter of thousands of young calves Australia-wide and some are seeking systems to convert low return 'bobby veal' to more profitable dairy beef.

Over the last two decades most state Department of Agriculture offices have published booklets on calf management and heifer rearing, such as in Queensland in 1983, NSW and Victoria in 1984 and Tasmania in 1991. To the authors of such booklets, I am grateful for the information they have collected and summarised. I would also like to acknowledge the many dairy farmers and advisers throughout Australia who have provided me with further data on the many aspects of calf rearing covered in this book. I trust I have painted a truly Australia-wide picture.

Prior to writing my first edition of this book, I made contact with a Canadian dairy researcher Dr Lumir Drevjany, based in Kemptville, Ontario. He produced a booklet on heavy calf production that was a valuable reference text when I wrote my pink veal farmer manual in 1990. More recently he has prepared an exhaustive list of methods by which calves can communicate with their rearers. He was kind enough to let me include some of his unpublished observations in Chapter 9 of this book.

Since the publication of the first edition of *Calf Rearing* (in 1993), considerable Australian research has been undertaken on improving existing systems of calf rearing, thus leading to an up-to-date second edition.

I would also like to thank my many departmental colleagues in northern Victoria for their encouragement and assistance in the preparation of both editions of this book.

John Moran, August 2002
Victorian Department of Natural Resources and Environment, Kyabram Dairy Centre
120 Cooma Rd, Kyabram, Vic. 3620, Australia
Telephone: +61 3 5852 0509
Fax: +61 3 5852 0599
Mobile: 0418 379 652
Email: john.moran@nre.vic.gov.au

Chemical warning

The registration and directions for use of chemicals can change over time. Before using a chemical or following any chemical recommendations, the user should ALWAYS check the uses prescribed on the label of the product to be used. If the product has not been recently produced, you should contact the place of purchase, or your local reseller, to check that the product and its uses are still registered. Users should note that the currently registered label should ALWAYS be used.

1

Introduction

Every year nearly five million beef calves and two million dairy calves are born throughout Australia.

Most beef calves are reared by their dams while farmers artificially rear some 800,000 calves for dairy heifer replacements and at least 100,000 for meat production. No data are available on actual calf losses, but a figure of 6–8% would not be unrealistic. This means that up to 60,000 calves die each year as a result of disease and/or poor management during early rearing. Heifer rearing has traditionally been considered a

Figure 1.1 Calf rearing is a science as well as an art

Figure 1.2 The 'dining room' in the calf shed must be kept clean and tidy

low input enterprise on many dairy farms and this contributes to high calf mortalities, poor growth rates to mating and low milk production of first-calf heifers.

Economic pressures are forcing dairy farmers to improve their farm productivity and this can be brought about through more intensive management practices. Increasing numbers of dairy farmers now consider better calf rearing as one avenue for improving farm efficiency. Up to one million week-old calves each year are slaughtered in Australia, making it one of the few developed countries in the world that still finds it more profitable to slaughter dairy bull and heifer calves, excess to herd requirements, for low value bobby veal rather than to grow them out for meat production. If this changes, dairy farmers will not be the only ones rearing young calves. Contract rearers, beef producers and even small holders could be selling weaned calves for growing out to dairy beef.

The principles of good calf rearing

This book describes the rearing of young calves up to 3 months of age, when they are at their most susceptible stage of life. They can then be run on pasture with no further need for milk feeding. The objective of good calf rearing is to produce healthy animals that will continue to grow into suitable heifer replacements for dairy herds, suitable dams for vealer herds, or suitable steers or heifers to grow out for eventual slaughter. Well-managed calf rearing should aim for:

Figure 1.3 Contract heifer rearing can be cost effective, whether in the feedlot or at pasture

1. Good animal performance with minimal losses from disease and death.
2. Optimum growth rate and feed efficiency.
3. Optimal cost inputs such as feed (milk, concentrates and roughage), animal health (veterinary fees and medicines) and other operating costs (milk feeding equipment, transport, bedding material, etc.) to achieve well-reared calves.
4. Minimum labour requirements.
5. Maximum utilisation of existing facilities such as sheds for rearing and pastures for grazing.

There is no single best way to rear calves, as all sorts of combinations of feeding, housing and husbandry can be successful in the right hands and on the right farm. Moreover, a system that works well on one farm may fail on another for reasons that are often inexplicable even to the expert.

By understanding the scientific principles of calf growth, nutrition, health and behaviour, producers can develop a system of husbandry that is successful on their own farm. If things go wrong, management can then be modified in a fundamentally sound way to put things right.

Rearing calves 'on the cheap' does not pay in the long run because a setback in early life cannot be compensated for later on.

This book does not aim to produce a recipe on managing young calves, as recipes are supposed to be foolproof and should work under all situations. Good calf management requires a certain degree of fundamental knowledge and empathy with the animals, but mostly it demands common sense.

The high costs of calf rearing

The first three months are probably the most expensive period in the life of cattle. During that time mortality rates can be very high, up to 8%, as is routinely recorded on US dairies. Australian farmers generally consider mortality rates of 2–4% as acceptable.

Concerted efforts must be made to ensure every calf is provided with and consumes sufficient high quality colostrum to provide the passive transfer of immunity to the many diseases that can inflict high losses (due to deaths and poor pre-weaning performance) during milk feeding.

With their undeveloped digestive tracts, calves require the highest quality and most easily digestible form of nutrients, namely whole milk or milk substitutes. Unfortunately, these are also the most expensive. As a source of energy, milk is four times more expensive than concentrates and 20 times more expensive than grazed pasture. This clearly shows that the most effective way of minimising the high feed costs of calf rearing is through early weaning and reduced milk feeding.

The need to protect young calves from the extremes of sun, wind and rain means that access to housing, or at least simple shedding, is essential during early life. With regard to the cost of maintaining healthy calves, most calf rearers could produce the bills for veterinary fees and drugs to show that disease management is more costly during these first three months of life. Unfortunately, many producers still consider it just too expensive to 'get the vet in' and, thus, depend on the calf's own defence mechanisms to fight off any disease. This results in needless calf deaths and suffering.

With increasing community concerns about animal welfare, calf rearers must be aware of and conform to codes of practice for the welfare of all calves they rear for replacement heifers and dairy beef, as well as for bobby calves – that is excess calves destined for slaughter within a week of birth.

For calf rearing to remain cost effective in the future, it will have to become more of a science than an art and producers will have to become more aware of the various costs involved. Using their own information on feed prices and feed quality, producers can calculate costs for their particular situation. To assist in this process, several commercial laboratories in Australia now test animal feeds for levels of dry matter, energy and protein.

An outline of this book

In providing background reading on many aspects of calf rearing, this book aims to give producers a better understanding of why certain management procedures work on their farm and why others do not.

Chapter 2 details the principles of digestion in calves. During early life this changes dramatically from a milk-only system to one that can digest both liquid and solid feeds, and finally to one that operates efficiently on grazed pasture alone.

Colostrum is the first milk produced by newly calved cows. Not only does it provide essential feed nutrients, it also supplies maternal antibodies that allow passive transfer of immunity against diseases of calfhood. Recommendations for colostrum feeding are discussed in Chapter 3.

Calf growth and development occurs not just because of the feed they eat but because that feed contains the essential ingredients of life, namely water, energy, protein, fibre, minerals and vitamins. These are discussed in detail in Chapter 4.

Chapters 5, 6, 7 and 8 outline much of what is involved in the obtaining of calves, their feeding of whole milk or powdered milk replacer and finally ration formulation of solid feeds. These chapters explain some of the diversity of management practices in detail.

All successful calf rearers develop an empathy with their animals. This comes about mainly through communicating with their animals to provide them with their day-to-day requirements. A Canadian colleague has listed over 60 different methods by which calves can communicate with humans – the ones most relevant to Australian systems are outlined in Chapter 9.

Chapter 10 explains the major diseases inflicting calves in Australia. The most widespread of these are scours, pneumonia and clostridia (which includes pulpy kidney, blackleg and tetanus). Most calves in Australia are run outdoors, but more and more farmers now use rearing sheds or hutches, particularly when handling large numbers of animals. Young calves still require some degree of protection against the extremes of sun, wind and rain, so all rearing systems should provide for housing during periods which may be stressful to their welfare and performance. This is discussed in Chapter 11.

Welfare aspects of calf rearing are discussed in Chapter 12. With a growing community awareness of the animal welfare issues of farming, there are an ever-increasing number of guidelines being introduced to control animal practices on farm.

Most artificially reared calves are destined for the dairy industry. The importance of post-weaning management has been highlighted in recent years as dairy farmers seek to improve the performance of their first-calf heifers. Target growth rates and other aspects of weaned heifer management are outlined in Chapter 13. This chapter also covers the role of dairy stock in beef production.

Like any other farm enterprise, calf rearing is a business and must be profitable. Many of the costs involved in rearing calves for milk or meat are outlined in Chapter 14. Budgets compare the total costs of different rearing systems.

Chapter 15 presents a checklist of best management practices (BMPs) for rearing dairy heifer replacements from birth through to first calving. These lists can be photocopied and placed on the notice board in the calf shed to remind rearers of the need to continually improve their young stock management as a key area in their overall farm management.

Appendices 1 to 3 summarise the 'Golden Rules' of calf and heifer rearing and explain many of the technical terms used in this book.

Texts for further reading

There are several good texts available on calf rearing, but these have been written for European and North American farmers where the emphasis is on high-cost indoor rearing with its greater potential for disease and its heavy labour demands. Fortunately, for much of the year, calves in Australia can be reared outdoors, where reduced disease and labour can minimise many of the traumas experienced by calves overseas.

Most Departments of Agriculture have booklets on rearing dairy heifer replacements. This book aims to provide the first single text that covers Australia-wide, with particular emphasis on the technical principles behind proven, successful management systems. It aims to be a useful reference for livestock producers, dairy and beef consultants, and also for students of animal science.

To provide additional reading, particularly for students, each chapter concludes with a relevant list of scientific references, many of which may only be obtainable from libraries in universities and Department of Agriculture institutes. Appendix 4 provides a list of key textbooks and bulletins on the principles and practices of calf rearing. It also lists relevant websites to guide readers through the maze of the Internet, to useful quality information on many aspects of calf rearing.

2

The principles of digestion of feed in calves

If all calves could be reared by their natural mothers, there would be little need for this book. Most beef cows do a good job of rearing their own offspring, provided due care is paid to their feeding and health and other aspects of their husbandry. The first essential of good husbandry in rearing calves is to keep them alive and fit enough to perform well later on. To do this, producers need to understand the development of the calf's digestive tract and the basic concepts of how calves digest their food.

The calf digestive tract

An adult animal needs four functional stomachs to give it the ability to utilise the wide range of feeds available.

The reticulum and the rumen harbour millions of microbes that ferment and digest plant material. The omasum allows for absorption of water from the gut contents. The abomasum, or fourth stomach, is the true stomach, comparable to that in humans, and allows for acid digestion of feeds.

The very young calf has not developed the capacity to digest pasture and so the abomasum is the only functional stomach at birth. Both newborn and adult animals have a functioning small intestine that allows for the alkaline digestion of feeds.

Figure 2.1 illustrates the anatomy of the stomachs and small intestine of a newborn calf. This schematic diagram shows the relative sizes of the four stomachs, the oesophageal groove, which runs from the oesophagus through the rumen to the abomasum, and the pyloric sphincter or valve at the bottom of the abomasum, which controls the rate of movement of gut contents into the duodenum.

The omasum and abomasum account for about 70% of the total stomach capacity in the newborn calf. By contrast, in the adult cow, they only make up 30% of the total stomach capacity (Figure 2.2).

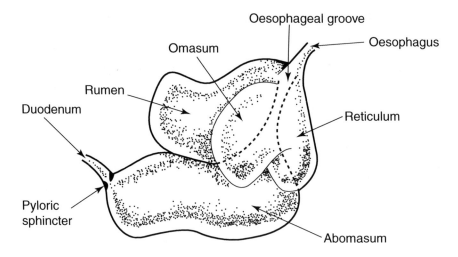

Figure 2.1 A schematic diagram of the four stomachs and duodenum of the newborn calf

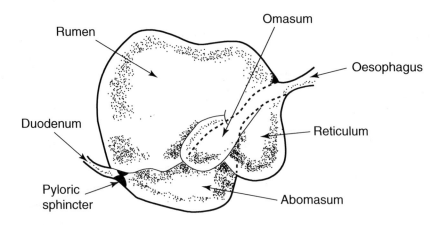

Figure 2.2 A schematic diagram of the four stomachs and duodenum of an adult cow

Digestion of feeds is aided by the secretion of certain chemicals called enzymes into the various parts of the gut. For example, calves produce the enzyme rennin in the abomasal wall to help digestion of milk proteins, while lactase is produced in the wall of the duodenum for digestion of milk sugar (lactose). These enzymes operate most effectively at different levels of acidity in the gut contents, acid in the abomasum and alkaline in the duodenum. To achieve this, the calf secretes electrolytes, or mineral salts, with the enzymes, to change the gut contents from one type to another.

The end products of digestion of the different components of feeds are absorbed through the gut wall into the blood stream where they are taken to the different parts of the body for the animal's growth and development.

The milk-fed calf

Milk or milk replacer, whether sucked from a teat or drunk from a bucket, is channelled from the oesophagus via the oesophageal groove into the abomasum. This groove is a small channel in the rumen wall that is controlled by muscles that allow liquids to bypass the rumen.

The groove is activated in response to different stimuli. It works well when calves suckle from their mothers or from teats, but sometimes does not work when they drink from a bucket. This appears to be a psychological condition in response to calves being separated from their mothers. Most calves can be trained by patient coaxing to drink quickly and well, responding to the new daily routine and the substitute mother in the shape of the calf rearer. When milk or milk replacer enters the abomasum, it forms a firm clot within a few minutes under the influence of the enzymes rennin and pepsin. This is the same process involved in making cheese or junket, using rennin to coagulate the milk protein. The clotting of milk slows down the rate at which it flows out of the abomasum, thus allowing for a steady release of feed nutrients throughout the gut and eventually into the blood stream. It can take as long as 12–18 hours for the milk curd to be fully digested.

The enzymes acting on milk proteins require an acidic environment and this is provided by hydrochloric acid secretion into the abomasum. However, until the acid digestion is operating efficiently, and this can take up to seven days, the only form of

Figure 2.3 Calves only require milk as their primary feed source for their first six weeks of life

protein that can be digested is casein. There is no substitute for casein in the very young calf. Milk replacers containing other forms of protein cannot be properly digested until the calves are older.

Digestion of milk can be improved by including rennet, which can be obtained from cheese factories, or commercial calf milk additives for the first week or so. These can provide additional acids to reduce abomasal pH and enzymes and specific bacteria to increase the rate of breakdown of the milk curd. Such additives are called probiotics, in that they aid in normal digestive processes. Research has not always found these to improve calf performance and health, and they are more likely to be beneficial when calves are suffering from ill health. Furthermore, their cost effectiveness has sometimes been queried.

Any milk from a previous feed is enveloped in this newly formed clot. Liquid whey protein and lactose are rapidly separated from the milk curd and pass into the abomasum. The milk fat embedded in the milk curd is broken down by another enzyme, lipase. This is secreted in the mouth in saliva and incorporated when milk is swallowed. Teat feeding rather than bucket feeding seems to produce more saliva and, hence, more lipase. Further digestion of the milk protein and fat occurs in the duodenum with the aid of enzymes produced in the pancreas.

Lactose, which is quickly released from the milk curd in the abomasum, is broken down to glucose and galactose and these are absorbed into the blood stream to form the major energy source for young calves.

Fats are broken down into fatty acids and glycerol for absorption and use as energy, while proteins are broken down into amino acids and peptides for absorption and use as sources of body protein.

Starch, from cereal grains, for example, is an important source of energy in older calves, but calves in their first few weeks of life cannot digest starch.

The abomasum is not acid until the calf is 1–2 days old and this has advantages and disadvantages. The major advantage is that the immune proteins in colostrum cannot be digested in the abomasum so are absorbed into the blood stream in the same form as when produced by the cow. This ensures their role as antibodies to protect against infection. The low acidity of the abomasal contents in the newborn calf constitutes a potential risk from the bacteria (and probably viruses) taken through the mouth. These will not be killed by acid digestion and they can pass into the intestines where they can do the most harm. All calves pick up bacteria in the first few days of life and this is essential for normal rumen development. However, the first bacteria to colonise the gut can also cause scouring. Provided the calf has drunk colostrum, the maternal antibodies can control the spread of these more harmful bacteria.

The milk-fed calf must then produce an acid digestion in the abomasum and an alkaline digestion in the duodenum. This is achieved by the production of electrolytes in the gut wall.

Calves suffering from scours due to nutritional disturbances or bacterial infections can lose large amounts of water and electrolytes in their faeces. These must be replenished as part of the treatment for scours.

Colostrum is the first milk produced by newly calved cows. Not only does it provide essential feed nutrients, it supplies maternal antibodies that allow passive transfer of immunity against diseases of calfhood. Recommendations for colostrum feeding are discussed in Chapter 3.

Rumen development and the process of weaning

When calves are weaned, the cost of rearing declines markedly. Feed costs are lower, labour inputs are reduced and incidence of ill health are less. It makes economic sense to wean calves as soon as is reasonable. However, the calf is forced to undergo several dramatic changes, namely:

- The primary source of nutrients changes from liquid to solid.
- The amount of dry matter the calf receives is reduced.
- The calf must adapt from a monogastric to a ruminant type of digestion, which includes fermentation of feeds.
- Changes in housing and management often occur around weaning, which can add to stress.

At birth, the rumen is a small and sterile part of the gut that by weaning must become the most important compartment of the four stomachs. It must increase in size, internal metabolic activity and external blood flow. The five requirements for ruminal development are:

- Establishment of bacteria.
- Liquid.
- Outflow of material (muscular action).
- Absorptive ability of the tissue.
- Substrate to allow bacterial growth, such as recycled minerals, as well as feed nutrients.

Prior to consumption of solid feeds, bacteria exist by fermenting ingested hair, bedding and milk that flows from the abomasum to the rumen. Most water entering the rumen comes from free water (actual water not water contained in milk or milk replacer solution). Milk will bypass the rumen via the oesophageal groove, whereas free water will not.

The rumen develops from a very small organ in newborn calves (1–2 L) into the most important part of the gut (25–30 L) by 3 months of age. It can enlarge very quickly during the first few weeks of life, given the right feeding management. Rumen growth only occurs under the influence of the end products of rumen digestion, which result from the fermentation of solid feeds by the rumen microbes. Development largely occurs through growth of rumen papillae on the rumen wall (leaf-like structures on the internal surface), which increase the surface area of the rumen and, hence, its ability to absorb these end products of digestion.

The rumen's capacity and the intake of solid feed are closely related. Rumen development is very slow in calves fed large quantities of milk. The milk satisfies their appetites so they will not be sufficiently hungry to eat any solid feed.

Rumination, or cudding, can occur at about 2 weeks of age and is a good indication that the rumen is developing. Solid feeds and rumination both stimulate saliva production and this supplies nutrients such as urea and sodium bicarbonate to produce the substrates for bacterial growth.

When early weaning calves, it is important to limit both the quantity of milk offered and its availability throughout the day. It is also essential to provide solid feeds. Roughages (low or high quality) should be offered in combination with high-quality concentrates. Roughages stimulate rumen development while concentrates supply feed nutrients not provided by the limited quantities of milk offered. Without the concentrates, calf growth is slow but the rumen still develops, resulting in undesirable pot-bellied animals.

Urea supplies nitrogen for the microbes, while sodium bicarbonate acts as a rumen buffer, helping to maintain a steady pH in the rumen contents. This is particularly important when calves eat large quantities of cereal grains later in life as the rumen microbes can produce a lot of lactic acid during fermentation.

Grain poisoning or acidosis occurs when lactic acid levels are excessively high and become toxic to the rumen microbes and eventually to the animal. As well as the end products that are absorbed through the rumen wall, microbial fermentation produces the gases carbon dioxide and methane and these are normally exhaled. When something prevents the escape of these gases from the rumen, bloat can result at any stage of life.

The role of roughage in the weaning process

There is continuing controversy about the role of roughage in the weaning process. Research in the 1980s clearly indicated that roughage was beneficial, whereas 1990s research found it was not always necessary. In most of the earlier research, calves were offered finely ground concentrates as pellets with or without long hay or straw. The inclusion of roughage in the diet improved intakes and performance and allowed for earlier weaning. In the later research, calves were generally offered the concentrates as coarsely ground meals plus some roughage, while some even included fine chopped roughages in the mix (sometimes called a muesli mix). In these studies, the inclusion of additional hay or straw was found to have little effect on pre-weaning performance.

Figure 2.4 The feeding of roughage to milk-fed calves is a controversial issue

Australian calf rearing systems often differ with those overseas, especially in seasonal calving areas. Far too many calves have to be reared at any one time to provide them all with individual pens for their entire milk feeding period. Consequently calves are reared in groups. Furthermore, the ingredients of most calf concentrate pellets are finely ground. In these situations, we have found clean straw to be a useful feed to include in the pre-weaning period. Some farmers prefer good quality hay, but these farmers usually have calves in very small groups, often one or two, so have greater control over roughage intakes.

It is difficult, and, hence, more expensive, for stock feed manufacturers to incorporate chopped hay into calf meals. Pellets are much easier as they will flow into silos. Therefore, dairy farmers should include the roughage component of the pre-weaning diet themselves.

Grazed pasture is not the ideal roughage source for milk-fed calves as it has too little fibre and a low feed energy density. Its high water content limits its ability to provide adequate feed energy for the rapidly growing animals. Until their rumen capacity is larger, young calves just cannot eat enough pasture unless it is very high in quality.

Calves reared on restricted milk plus concentrates display good rumen function by 3 weeks of age and have sufficient rumen capacity for weaning by 4–6 weeks of age. However, if the diet was restricted milk plus high quality pasture, rumen capacity may not be sufficient for weaning until 8–10 weeks of age. Even then, growth rates would be lower in calves weaned onto pasture alone because of insufficient energy intake due to the physical limitations of rumen capacity.

If too high in quality and fed ad lib (that is, fed to appetite), calves will prefer the roughage to concentrates, leading to a reduced intake of feed nutrients and slower growth. When clean cereal straw and concentrates are both fed ad lib, together with limited milk, calves will eat about 10% straw and 90% concentrates. Without the roughage and the resulting rumination, rumen development will be slower due to insufficient saliva and end products of fibre digestion.

References and further reading

Thicket, B., Mitchell, D. and Hallows, B. (1988), *Calf Rearing*, Farming Press, Ipswich, England.

Quigley, J. (1997), 'Evaluating and Optimising Calf Performance', *Proc. First National, Professional Dairy Heifers Growers*, Atlanta, Georgia.

Webster, J. (1984), *Calf Husbandry, Health and Welfare*, Granada, Sydney.

3

The importance of colostrum to newborn calves

Calves are born with no immunity against disease. Until they can develop their own natural ability to resist disease, through exposure to the disease organisms, they depend entirely on the passive immunity acquired by drinking colostrum from their dam.

Colostrum is the thick, creamy yellow, sticky milk first produced by cows initially following calving, and contains the antibodies necessary to transfer immunity onto their calves. Also called 'beastings', it is essentially milk reinforced with blood proteins and vitamins. It has more than twice the level of total solids than in whole milk through boosted levels of protein and electrolytes. It also contains a chemical allowing newborn calves to utilise their own fat reserves to immediately provide additional energy.

The concentrations of protein, vitamins A, D and E in colostrum are initially about five times those of whole milk, with a protein content of 17–18% compared to 2.5–3.5%. However, within two days these are little different from those in whole milk. The levels of vitamins in colostrum are dependent on the vitamin status of the cow. The blood proteins transfer passive immunity from mother to offspring through maternal antibodies or immunoglobulins (Ig).

The chances of calves surviving the first few weeks of life are greatly reduced if they do not ingest and absorb these antibodies into their blood stream. It takes far less disease organisms to cause disease outbreaks in such calves than if they can acquire immunity from their dam. Calves without adequate passive immunity are up to four times more likely to die and twice as likely to suffer disease than those with it. Furthermore, in certain situations, blood levels of antibodies in heifer calves are directly related to their milk production in later life.

The term colostrum is generally used to describe all the milk produced by cows up to five days after calving, until it is acceptable for use by milk factories. However, a more correct term for milk produced after the second milking post-calving is transition milk. This milk no longer contains enough Ig to provide maximum immunity to calves, but still contains other components that reduce its suitability for milk processing. Milk factories can now test for and penalise farmers who include transition milk in their milk vat. As it has no market value, transition milk should be fed to calves to reduce their total feed costs. However, it must be stressed that the immune properties of this pooled milk are much reduced once first milking colostrum is diluted with that from second or later milkings.

When considering colostrum feeding to dairy calves, it should be appreciated that modern milking cows are vastly different to the primitive, feral cows from which they evolved thousands of years ago. Their udders are much larger and often hang too low for easy suckling by their offspring. They produce vastly greater quantities of milk, which means that their first and second milking colostrum is much more diluted than is desirable for optimum quality. Furthermore, as mothering ability has little relevance on dairy farms and has probably been bred out of cows, they are less likely to want to suckle their progeny immediately after birth. This is still not the case with beef cows, where unassisted suckling is a highly efficient means of passive transfer of immunity in beef calves. These natural methods are less effective in dairy herds, meaning that farmers often have to rely on so-called less natural techniques.

Changes in recommendations on colostrum feeding

Recommendations for colostrum feeding have changed dramatically over the last decade. Ten years ago, it was considered acceptable for all calves to run with their dams for one, two or even three days and for her to pass on passive immunity through natural suckling. As producers learnt more about the causes and prevention of calf diseases, they became more 'colostrum conscious'. Current advice to farmers is to ensure calves drink from their dam within the first three to six hours of life, and if not then to provide additional colostrum from its mother or another freshly calved cow. Colostrum quality can be assessed visually or using a colostrometer, which works on the same principle as the hydrometer used to measure the acid level in car batteries. Recently, more sensitive field test kits have become available to Australian calf rearers.

Two feedings during the first day, 6–12 hours apart, and each of 2 L (litres) of good quality colostrum used to be considered sufficient to provide passive immunity, mainly because of concern about the small capacity of the abomasum in newborn calves. However, some overseas advisers now recommend that dairy farmers remove the calf as soon as possible after birth (within 15 minutes) and feed it 3–4 L of top quality colostrum at one feeding. This can be via teat, bucket or stomach tube. The latest findings are that this extra colostrum will be stored in the rumen, from where it slowly passes through the abomasum into the intestines where the Ig are absorbed into the blood.

The importance of colostrum to newborn calves | 17

This chapter highlights the important principles behind colostrum feeding to ensure that all calves get a good start to life through adequate transfer of passive immunity. These principles can be categorised into three Qs: quality, quantity and quickly:

- Quality is providing good quality colostrum.
- Quantity is ensuring calves ingest sufficient antibodies.
- Quickly is timing the first feed to ensure efficient absorption of the antibodies into the blood.

Another way US farmers remember the essentials of colostrum feeding is as ABC:

- As soon as possible.
- Best quality.
- Chug a full gallon.

Colostrum quality

Newborn calves need to ingest at least 100 g of Ig within their first three–six hours of life, and ideally the same amount 12 hours later. The quality of colostrum is expressed in terms of its Ig concentration, with excellent quality colostrum containing at least 90 g/L, good quality (65–90 g/L), moderate quality (40–65 g/L) and poor quality (less

Figure 3.1 Colostrum quality must be assessed to ensure immunity can be transferred to calves

than 40 g/L). The volume of colostrum that should be drunk to supply 100 g of Ig can then be calculated from its quality.

The higher the colostrum quality, the faster and more efficiently the Ig are absorbed by newborn calves. With poor quality colostrum, not only must calves be fed very large volumes to provide sufficient Ig intakes, but it is likely that even then, inadequate amounts of Ig will be absorbed into the blood. For example, 2 L of colostrum containing 80 g/L of Ig will provide more passive immunity than 4 L of colostrum containing only 40 g/L of Ig.

After their first milking, dairy cows begin to reabsorb the Ig back into their udder tissue. For this reason, colostrum from the second milking contains only half the Ig content as that from the first milking. Cows are generally deficient in Ig levels if they have been previously milked, or are seen to be leaking milk, prior to calving. Colostrum quality is also low in induced cows or those with less than four weeks between drying off and calving. Colostrum quality does not seem to be affected by pre-calving feeding management.

Older cows, and cows raised in the herd (compared to those purchased as in-calf heifers), will generally produce better quality colostrum, containing more antibodies for diseases existing on that farm. First-calf heifers are likely to have the lowest levels of antibodies in their colostrum because they have had less exposure to these diseases. Bloody colostrum may also be lower in antibody levels.

On the whole, Jerseys produce colostrum containing more Ig than do Friesians. In fact, very few Friesian cows produce excellent quality colostrum. Cows yielding large volumes of first milking colostrum (8 L or more) are more likely to have low Ig levels. There is little seasonal effect on colostrum quality. Some studies have found seasonal differences in acquired immunity in calves, but this is more related to changes in colostrum feeding management rather than colostrum quality.

Two recent studies in Victoria have confirmed US findings that many of our dairy cows do not produce good quality first milking colostrum. Using a colostrometer to assess quality, colostrum was categorised as above, with the percentages of cows in the two studies producing poor (43 and 40%), moderate (31 and 37%), good (23 and 19%) and excellent (4 and 4%). These findings highlight the importance of identifying cows producing poor quality colostrum soon after parturition then feeding their calves on stored good or excellent quality colostrum (Humphris 1998).

The immune properties of colostrum can be enhanced by vaccinating cows at drying off. There are vaccines to improve calf immunity against E. coli, clostridia, leptospirosis and salmonella. Selection of the most appropriate vaccines should be based on the prevalence of particular calf diseases in the area, information readily available from local advisers and veterinarians. American farmers are fortunate in having vaccines against two other major causes of calf scours, rotavirus and cryptosporidia, and also against several respiratory infections. As demand for better quality colostrum increases, such dry cow vaccines should become available in Australia.

It is possible for Johne's disease to be transmitted from cow to calf prior to calving, and also via the colostrum. Although the incidence of transfer by these methods may be quite small, they should be taken into account in any Johne's disease prevention program. For this reason, current recommendations are for disposal of both the infected cow and her daughter, if the last calf before breakdown was a heifer. The most important aspect of any Johne's disease control program is to minimise contact between milking cows and replacement heifers until at least 6–12 months of age. Rearing systems based on milk replacers rather than whole milk are often recommended, but this is mainly to reduce the possibility of cross-infection from cows being milked in the dairy to calves being fed fresh whole milk in the rearing shed or paddock.

In well-managed herds with few disease challenges, calf isolation is particularly important as the colostrum is likely to have lower Ig levels. Providing calves with both isolation and high quality colostrum is also important in herds with high culling rates, as they would contain a higher proportion of first-calf heifers. Farmers with seasonal-calving herds and practicing oestrus synchronisation of heifers to calve before the older cows face a quandary of not having fresh supplies of high quality colostrum on hand. These farmers may consider storing colostrum from the previous year, which means they must be able to identify the best quality colostrum for long-term storage. If concerned about the quality provided by any freshly calving cows, then additional colostrum should be fed to their calves from a store of good quality colostrum.

Identifying and storing good quality colostrum

Over the years, there have been various attempts to assess the quality of colostrum. Unfortunately, visual appraisal is a poor way of assessing quality because thick, creamy colostrum may simply be indicative of its high fat content. There is a negative relationship between Ig level and fat content in colostrum as the Ig reside in the non-fat component of the milk solids. The colostrometer was developed specifically to determine Ig levels, but recent research has shown up some limitations. When using colostrometers to quantify Ig status, it is important to measure it at room temperature or, better still, using a thermometer to make certain the colostrum is measured at the recommended temperature. Colostrometers are good at detecting poor samples, but unfortunately they are limited in their ability to make other assessments. Their major role should be to screen samples to ensure only colostrum with more than 80 g/L of Ig is fed fresh or stored for later use.

Colostrum allowed to sour or become overheated loses its antibody effectiveness. Freezing is the best method for storage, as this will retain its antibody activity for at least one or two years. Frost-free freezers are not ideal as their freeze-thaw cycles can allow the colostrum to thaw, and this can shorten its effective storage life. Colostrum can also be stored for up to 10 days in the refrigerator. Be sure to identify containers of stored colostrum with its origin and, if known, its quality.

Frozen colostrum should be thawed out carefully, as overheating it above 50°C denatures the Ig. Colostrum frozen in 1–2 L milk cartons or, better still, in thick plastic bags can be thawed out in 50°C water. The time between the first appearance of calf's feet prior to birth until it is first ready to drink, should be sufficient for frozen colostrum to thaw in warm water at 50°C. If using a microwave oven, it should have a turning tray to avoid hot spots and the defrost setting should be used. Pour the liquid off as it thaws to reduce overheating.

Feeding colostrum to newborn calves

US veterinarians now recommend feeding 4 L rather than 2 L of good quality colostrum at first feeding, just to ensure that adequate Ig are ingested. Increasing volumes at first feeding markedly reduces the number of calves with low blood Ig levels, indicative of failure of passive transfer of immunity. It takes the first 2 L to fill the rumen while the second 2 L spills over into the abomasum. Newborn calves should readily drink 2 L through a teat from a nipple bottle, however, greater volumes are generally refused.

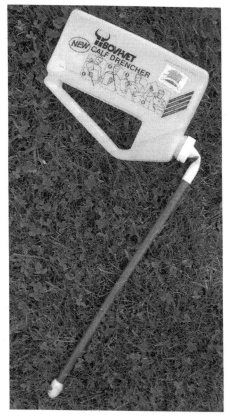

Therefore, stomach tubing is necessary to ensure the entire volume is consumed. Fluids will pass directly into the rumen, not the abomasum, because the oesophageal groove will not close. However, it will quickly flow from the rumen into the abomasum. Farmers should always have a stomach tube on hand and learn how to use it. Calves weak from difficult calvings or with swollen tongues preventing them from sucking should be tube fed the entire 4 L of colostrum immediately. Calves will not regurgitate it or get it into their lungs if the fluid is correctly administered with a stomach tube. Veterinarians are not overly concerned about providing the calf with too much colostrum and causing scours, because this first milk contains less lactose than later

Figure 3.2 Ideally, colostrum should be given by stomach tube to ensure every calf consumes sufficient volume

milk, and it is the overflow of lactose into the hind gut that leads to scours. The first drink is the most important as the ability of calves to absorb further Ig through the gut wall drops off markedly thereafter.

How to stomach tube a calf

Weak calves may not be able to drink liquids from a teat. Stomach tubing is the best way of ensuring that they consume enough liquid. Some dairy farmers routinely stomach tube all newborn calves to provide colostrum and be certain that they will absorb sufficient antibodies to enhance their immunity against diseases. Scouring calves with severe dehydration, that are too weak to drink themselves, can also be stomach tubed.

The stomach tube is a flexible piece of plastic tubing with a tear-shaped end, which is designed to be easily inserted into the oesophagus, but not into the lungs. It is usually attached to a plastic container holding the liquid to be fed.

The first step in using the stomach tube is to determine the length of tube to be inserted. This is measured as the distance from the tip of the calf's nose to the point of its elbow behind the front leg, usually 45 cm or more. This point can be marked on the tube with a piece of tape.

Ideally, the calf should be standing so the fluids are less likely to back up and enter its lungs. However, calves that are too weak to stand can be tubed in a sitting position and even while lying down. The stomach tube is easier to use when calves are restrained. Young calves can be backed into a corner for better head control. A calf allowed to throw its head from side to side may injure itself or you.

If the weather is cold, the tube can be placed in warm water to make it more pliable. The tube should be dipped into a lubricant, such as mineral or vegetable oil. The tip of the tube is then placed into colostrum or whole milk, whichever is to be fed. Calves may suck the end of the tube, making it easier for it to pass into the oesophagus. A calf's mouth can be opened by gently squeezing the corner of the mouth or by grabbing its head over the bridge of the nose and gently squeezing the upper palate or gums.

Once it is opened, the empty tube should be passed slowly along the tongue to the back of the mouth. When the tube is over the back of the tongue, the calf starts chewing and swallowing it, after which the tube is passed down into the oesophagus. The end of the tube can be felt quite easily. Never force the tube; if it is being correctly put down the oesophagus, it should slide in quite easily.

After the tube is in place and before any fluids are given, it should be checked for proper positioning in the oesophagus. If it is properly positioned, the rings of the trachea (leading into the lungs) and the rigid enlarged oesophagus can be felt easily. If you cannot feel both of these, remove the tube and start again. The exposed end of the tube should be checked for spurts of air, which indicate that the tube has gone into the lungs. The calf will often cough, but not always, if this occurs.

The tube can be unclipped or straightened out, or the container can be tipped up to allow liquid to flow down into the stomach. Liquids should be at body temperature (38°C) in order to prevent shock to an already weak calf. It may take three minutes or more to allow sufficient fluid to be administered. The calf will regurgitate less with a slow flow rate.

When feeding is over, the tube should be slowly removed. The tube should be cleaned and sanitised, then allowed to drain and dry.

Veterinarians often concede that this method for colostrum feeding is not natural, but it does provide an easy and well-tolerated method of achieving an adequate and early intake of Ig at the first feeding. It is widely used overseas with increasing application in Australia, where calf rearers are finding significant improvements in passive transfer of immunity and reduced calf health problems.

Using stomach tubes to relieve abdominal pressure

Stomach tubing can also be used to relieve pressure build up in the stomach during milk bloat. This can occur with twice daily feeding of some milk replacers. This is the result of the previous curd of clotted milk not being given sufficient time to digest before the calf is offered further milk replacer.

Milk bloat can also occur when milk enters the rumen rather than the abomasum and ferments with other ruminal contents. This can restrict the escape of ruminal gases. This may occur when calves have functional rumens and, in these cases, they can be weaned off milk. Where this problem occurs in immature calves, a veterinarian should examine them further.

To relieve pressure build up, introduce the tube without the bottle attached, into the calf's oesophagus as described above. The gas should then be heard escaping, generally with a foul smell, and the distended abdomen will quickly return to normal.

Commercial aids to calf rearing

Colostrum or electrolytes can be administered to sick or weak calves using the McGrath Fluid Feeder ($38), marketed in Australia by Heriot Agvet, Rowville, Victoria, phone (03) 9764 9588. This is a collapsible 2 L bottle with a teat or 50 cm long stomach tube. It is preferable for the calf to suckle but if it can't, the stomach tube can be easily and safely slid into the oesophagus by following the instructions. A store of colostrum should be kept frozen in 2 L plastic bottles. See Chapter 6 for further details on colostrum storage.

Another excellent cheap stomach tube is the Bovivet Calf Drencher ($30), a Danish product marketed by Shoof International, Private Bag 522, Cambridge, New Zealand, phone (07) 827 3902, fax (07) 827 7596. This is a hard plastic 2 L bottle with a handle and a tube for dosing. The handle allows the calf to be firmly held with one hand and the drencher with the other. There is a soft latex bulb at the end of the tube to ensure it cannot go down the calf's larynx (wind-pipe).

Calf Guard colostrometers for testing the Ig quality of colostrum were manufactured by NorthField Laboratories (now Numico Research), Adelaide, but because of reduced demand, are no longer available. A New Zealand model (manufactured by Shoof International) is now available for $100 from Australian Calf Rearing Research Centre, PO Box 54, Heyfield, 3858, Victoria, phone (03) 5148 9189.

It is one thing offering a sufficient volume of high quality colostrum and another ensuring that the calf actually absorbs the Ig. For many years, calf rearers in the US and Europe have been able to assess the quality of colostrum and the immunity status of calves. Such field test kits are now commercially available in Australia, imported by Quick Test Distributors (Australia), PO Box 547, Mascot, 2109, NSW, phone 1800 147 475, fax (02) 9384 8922, email sales@midlanz.com, website www.midlanz.com. Three Midland Quick Test Kits are also available that do not require any laboratory equipment or a set temperature, these being:

1. Bovine colostrum IgG Kit, which measures the amount of immunoglobulin G (IgG) in bovine colostrum, with results obtained within 20 minutes. Cost is $13/calf.

2. Whole blood IgG Kit, which measures IgG in whole blood of newborn calves, together with a blood anticoagulant, provides results within 20 minutes. Cost is $12/calf.

3. Calf plasma IgG kit, which measures IgG in newborn calves, following clotting of the blood, with results obtained within 10 minutes. Cost is $12/calf.

There is also a second blood test kit for newborn calves available to calf rearers in Australia. Based on a glutaraldehyde test to detect Ig levels in calf blood, it is called Gamma Check B ($6) and is imported from the US by MAVLAB, in Slacks Creek, Queensland, phone (07) 3808 1399.

Many of the above products are also available from a newly established Gippsland centre, Australian Calf Rearing Research Centre, PO Box 54, Heyfield, 3858, Victoria, phone (03) 5148 9189, website www.australiancalfrearingresearchcentre.com (also www.acrrc.com). They are also the Australian agents for all Shoof (NZ) products.

Valuable newborn calves with low levels can be given additional Ig by injection, during the first four days after calving; whether these provide much improved immunity depends on the range of infectious organisms they will encounter during rearing. If bought-in calves have low levels of Ig, it is too late for them to obtain this from frozen colostrum stored on-farm, but at least these calves are identified as being more susceptible to disease.

Results from overseas research on colostrum feeding

A large-scale survey of 600 dairy farms throughout the US (NAHMS 1994) found that more than 40% of newborn calves had immunity levels below those recommended, while 25% of these calves had critically low immunity status. The death rate amongst all these calves was twice that of calves with adequate passive transfer of immunity. The study concluded that over 20% of the calf deaths in the US could be avoided by ensuring adequate and timely colostrum intakes. Australian surveys confirm these findings.

This study allowed a comparison of the effects of varying blood Ig levels on calf performance for 2020 calves, presented in Table 3.1. As blood Ig levels increased, calves grew faster, had more efficient feed utilisation, had lower incidences of scours, but of most importance, had much lower mortality rates. The extremes of mortality were 29% in the 6% of calves with very low blood Ig levels (0–5 mg/mL) compared to only 8% in the 66% of calves with good blood Ig levels (>15 mg/mL).

Table 3.1 Four-week performance of calves with varying levels of blood Ig levels

Blood Ig level (mg/mL)	0–5	5–10	10–15	15–25	>25
% of calves	6.0	11.0	16.0	29.0	37.0
4-week live weight gain (kg)	9.6	10.7	11.0	11.1	11.6
Feed conversion (kg feed/kg gain)	2.7	2.1	2.2	2.0	1.8
Average faecal score	1.4	1.3	1.2	1.2	1.2
Scour days	7.3	5.7	4.8	5.1	4.9
Mortality (%)	29.0	16.0	11.0	8.0	8.0

Faecal scores: 1, normal; 2, loose; 3, watery; 4, blood or mucous
Scour day, any day when faecal score is 2 or more

If calves are left to nurse from their dam for 24 hours, more than 60% do not take in sufficient Ig. In this US study, a second group of calves were provided with 2 L of good quality pooled colostrum as soon as possible after birth, and they recorded a 19% failure of passive transfer of immunity. A third group fed 3 L of colostrum had a 10% failure of passive transfer.

Managers of herds with high yielding cows differed in their newborn calf care to those with lower yielding cows. Top producers were more likely to separate calves from dams at birth before nursing, feed colostrum either by bucket or stomach tube, and feed 4 L/calf or more. Bucket feeding the colostrum was more popular than using stomach tubes although more used stomach tubes in larger herds. Among farmers that allowed calves to nurse their calves, top producers were more likely to supplement colostrum delivery with hand feeding.

As a result of findings such as these, two thirds of US dairy farmers now artificially feed the first colostrum to their calves. Of those still allowing newborn calves to nurse from their dam, 40% assist each calf, thereby increasing their chances of receiving adequate and timely quantities of first colostrum. Although providing calves with immediate assistance to stand and suckle increases the ingestion of Ig, many calves still show failure of passive transfer of immunity.

Delayed sucking is the major cause of poor transfer of immunity as overseas surveys have shown that 25–30% of calves fail to suckle by six hours and nearly 20% by 18 hours. The first suckling is later in calves born to heifers and in those born to cows with low, pendulous udders. Diseases are more prevalent in calves that have delayed their first suckling.

Furthermore, there is considerable variation in the actual quantities of colostrum drunk by naturally suckled calves, but the average intake is only 2.5 L within their first 24 hours of life. Unless this is very high quality colostrum, inadequate Ig intakes will occur.

These findings demonstrate several important aspects of colostrum management. The very high failure rate with calves nursed by their dams is due to the inability of calves to drink sufficient colostrum within a few hours after birth. The udder of the cow at birth is large and at times painful, making drinking of sufficient colostrum difficult. Furthermore, calves born weak or having difficult births may not even stand for several hours. As already mentioned, the Ig concentration of the colostrum may be low, meaning that calves have to voluntarily drink large volumes at a time when both cow and calf would prefer to rest.

Because the gut absorbs less Ig following this first drink, it is preferable to prevent the calf from suckling her dam unless the colostrum is guaranteed to be high quality. There will also be room in the calf's stomach for the administration of additional selected colostrum.

Calves fed by teat, bucket or stomach tube absorb Ig with equal efficiency. The presence of its dam during artificial feeding can improve the Ig absorption. The colostrum could be warmed to body temperature, so that calves will not require additional body energy during its digestion. Shivering of calves after drinking cold fluids, common in cold and wet weather, will thus be prevented.

Recent research in Australia, at Flaxley in South Australia, is assessing the variation in absorption of colostral Ig (Neville, personal communication) that can vary from 15–60% in calves born at the same time. This may be linked to the calves' ability to clot the colostrum in the abomasum. If it fails to clot, it passes to the small intestine where absorption is less efficient.

The team found that 43% of newborn calves failed to form a clot from 2 L of milk within 1.5 hours of feeding. Three factors could reduce clotting ability in the abomasum. Firstly, colostrum does not clot as readily as normal milk. Secondly, newborn calves produce rennin of lower clotting ability of older calves. Thirdly, amniotic fluid from the dam is usually present in the abomasum, which might inhibit rennin activity.

The addition of rennet improved clotting ability and Ig absorption, so the team hopes to develop a suitable rennet tablet to aid clotting and, hence, passive transfer of immunity in newborn calves.

Overseas farmers have access to artificial colostrum, which can be used to supplement that from freshly calved cows. These are becoming available in Australia, but it must be remembered that they are only supplements and should not be used to replace good quality colostrum.

Timeliness of colostrum feeding

Every half hour after birth that colostrum feeding is delayed, antibody transfer decreases by 5%. A calf that does not drink until 6 hours old has already lost the opportunity for 30% of the possible antibodies entering its bloodstream.

Colostrum feeding can then be seen as a race between the arrival of the protective colostrum Ig in the calf's intestines and the disease-causing pathogens. The longer calves are without Ig, the more opportunity for these pathogens invade the gut. If certain pathogens, such as E. coli, 'win the race' in the first few hours, they can even be absorbed into the blood, causing severe scours and reducing the effectiveness of any absorbed Ig.

Time of colostrum feeding is crucial. The cells in the intestinal wall mature in these first 12 hours, eventually shutting down their absorption mechanism. Furthermore, after 24 hours, when the abomasum starts to secrete acids to make the milk-digestive enzymes more effective, these degrade the Ig proteins, which reduces their effectiveness. Another confounding factor is that protection against pathogenic bacteria is minimal until the abomasum can secrete sufficient acid to reduce their potential to cause scours. Provided the calf has drunk colostrum, the maternal antibodies can control the spread of these harmful bacteria. The Ig in colostrum are still beneficial to the calf even if it can no longer be absorbed into the blood, as they line the intestinal walls to provide local protection against the build up of pathogens.

Stressed calves, such as those born in cold, rainy weather and left unprotected, and those requiring assistance during birth, cannot absorb Ig for as long a time as calves with easier births. Calf rearers who feed colostrum by artificial means give these calves a greater chance of survival.

The current US recommendations for first feeding are to offer 3 L to Jerseys and small Friesians and 4 L to average size Friesians, as soon as possible after birth. A second feeding of 2 L, or more if the calf is willing, can be provided six hours later, although it is not really necessary to feed these calves until the following day. Colostrum and transition milk should then be fed for the first few days of life to provide local protection against disease.

If calves are not weak from a lengthy and difficult birth and are breathing well, they do not require further stimulation from their dams such as licking off. There is little benefit for cows to lick their calves and strengthen maternal bonds. In searching for the teat, calves are likely to take in more pathogens than they would get from a sterilised stomach tube.

The longer calves spend with their dams, the greater their chances of contracting disease. The practice of 'snatch calving', or removing the calves from their dams at birth, may be difficult to encourage in seasonal calving herds unless there are obvious benefits through reduced health problems, such as Johne's disease, and reduced mortalities. It would greatly increase labour requirements during the busy calving period.

Summarising good colostrum feeding management

Year-round calving herds

In summary, the important principles of good colostrum management in year-round calving herds are:

1. Do not use colostrum from mature cows that produce more than 8 L at their first milking.
2. Use only first milking colostrum.
3. Feed 4 L to large calves or 3 L to smaller calves at first feeding.
4. Feed colostrum as soon as possible, at least within the first three hours after birth.
5. Do not let calves suckle their dams.

Seasonal-calving herds

How many of these recommendations should be followed in seasonal-calving herds depends on the increased labour requirements and their effectiveness in improving disease resistance in young calves. Certainly, colostrum quality should be more closely monitored than it is currently, and calves should be separated from their dams within the first three–six hours. Calves born to very highly yielding cows or cows with very large pendulous udders are more likely to require supplemental colostrum.

Any producer with major disease problems during calf rearing, or a high calf mortality, say more than 2–3%, should seriously consider blood testing several healthy calves to quantify their immunity status. If the majority of calf mortalities and/or the poor 'doers' are from animals born to first-calf heifers, these animals should be routinely artificially fed their first feed from a frozen colostrum bank or from selected colostrum derived from older animals.

US recommendations

A US organisation, the Bovine Alliance on Management and Nutrition (BAMN), has developed a series of farmer guidelines for calf management with one specifically on colostrum (BAMN 1995). Table 3.2 illustrates their colostrum management recommendations.

Table 3.2 Current recommendations for post-calving management of dairy herds in the United States

Calving area	– Ensure cows calve on a clean, calving pad or on a clean pasture
Separating calves	– Separate calf from dam as soon as possible
Colostrum feeding	– Feed first feeding of colostrum as soon as possible (within one hour) – Use fresh colostrum from the dam, if good quality – Feed at least 3 L in the first feeding and again 12 hours later – if colostrum is assessed as good quality, feed 2 L at first feeding – Feed 3 L at each feeding if the calf weighs more than 54 kg, has not consumed colostrum within the first six hours, or if the calving area is dirty – Use an oesophageal feeder if the calf will not consume sufficient colostrum
Colostrum quality	– Measure colostrum quality with a colostrometer before use – Use only good quality colostrum – Save good quality colostrum by freezing in 1 or 2 L plastic bottles – Use fair to poor quality colostrum and transition milk only for older calves
Other management tasks	– Dip navels in or spray them with tincture of iodine as soon as possible – Put calf in an isolated, dry and draft-free environment – Continue to feed lower quality colostrum or transition milk for two or three days after birth

Farmers who do not practice good colostrum feeding management may find their calves still remain healthy and grow well. Certainly, some calves with low blood Ig levels are healthy and productive. This reflects the importance of other aspects of calf rearing, such as good hygiene, lack of cold stress, empathy from rearers and good milk feeding management. However, as our expectations for good pre-weaning calf performance increase, the importance of improving immunity against disease becomes paramount. It is highly unlikely that dairy farmers can develop a calf rearing system with minimal health problems and that routinely produces weaned heifer replacements weighing 100 kg at 12 weeks of age without a sound colostrum feeding program.

Value-adding colostrum

Up until several years ago, colostrum had little market value, unless calf rearers required more to supplement their own supplies. This is changing as medical and sports scientists document the therapeutic potential of colostral antibodies on human health and performance. Bovine colostrum is now being collected from dairy farmers in Victoria, freeze dried (so as not to denature the Ig) and used for reducing diarrhoea in babies and enhancing the performance of elite athletes.

Colostrum from the first and second post-calving milkings is bulked and tested for Ig status using a colostrometer. If of sufficient quality, the colostrum can return up to $2/L, compared to 28–30c/L for vat milk (Galt 2000).

This new market opportunity provides farmers with two benefits. Firstly, it can greatly increase returns from milk that previously had no market value. Secondly, by quantifying the Ig status of the colostrum, it allows farmers to more objectively decide on the level of colostrum feeding for their replacement heifer calves. In so doing, it will improve calf immunity levels, thus reaping the benefits as described below.

Financial benefits from good colostrum feeding practices

Data from a large-scale US calf rearing unit allowed for the calculation of the financial benefits arising from optimum colostrum management (Fowler 1999). When comparing performance of 335 calves with low immunity (0–9.9 mg/mL of Ig) to those of 1663 calves with high immunity (>10 mg/mL of Ig), there were four major benefits, as follows:

1. **Calf weight gain.** Low immunity calves gained 10.3 kg compared to 11.3 kg in high immunity calves. Valuing each kilogram live weight gain at US$1.50, each high immunity calf returned US$1.50 more.

2. **Feed conversion.** Low immunity calves required 2.35 kg feed/kg live weight gain compared to only 1.95 kg feed/kg live weight gain in high immunity calves. Over four weeks, the high immunity calves consumed 5.4 kg less feed, when valued at US$1.05, represented a saving of US$5.70/calf.

3. **Incidence of scours.** Low immunity calves had 6.3 scour days compared to only 4.9 scour days in high immunity calves. Costs of antibiotics and electrolytes were US$10.80 for low immunity and US$7.10 for high immunity calves, a difference of US$3.70. These costings did not take into account any additional veterinary bills to treat the scouring calves.

4. **Calf deaths.** Mortality rates were 20.7% in low immunity calves, compared to only 8.6% in high immunity calves. Valuing each calf at, say, US$100 each, savings were US$12.10 per high immunity calf.

The total savings arising from optimum colostrum feeding practices then amounted to US$23, or nearly a quarter the value of each replacement heifer calf.

To directly transfer these results to Australia, I have assumed that calf rearing costs are in the same relative proportion in Australia as they are in the US. Furthermore, I have costed the semen and artificial insemination (AI) required to produce one live heifer calf to be at least A$100. Therefore, dairy farmers could save A$23/calf through ensuring all calves are given the opportunity to develop high levels of immunity following birth.

References and further reading

Bovine Alliance on Management and Nutrition (1995), *A Guide to Colostrum and Colostrum Management for Dairy Calves*, Arlington, Virginia, US.

Fowler, M. (1999), 'What is it Worth to Know a Calf's Ig Level?', *Proc. 3rd Prof. Dairy Heifer Growers Ass.*, p.31–6, Bloomington, Minnesota.

Galt, D. (2000), 'Production and Sale of Colostrum: Is it a Viable Proposition?', *Asian-Aus. J. Anim. Sci. 13 Suppl July 2000*, A:316–7.

Garry, F. (1995), 'Enhancing Dairy Calf Health: How Colostrum Works and How to Optimise Its Use', *Proc. Aust. Ass. Cattle. Vet*, p.117–21, Melbourne.

Humphris, T. (1998), 'The Effect of Natural Suckling, Compared with Natural Suckling and Artificial Supplementation with Colostrum, on the Level of Passive Transfer of Immunoglobulins in Calves', *Proc. XX Wld. Assoc. Buiatrics Cong.*, p.345–50, Sydney.

National Animal Health Monitoring System (1994), *Dairy Heifer Morbidity, Mortality, and Health Management, Focusing on Preweaned Heifers*, USDA, Center for Epidemiology and Animal Health, Fort Collins, Colorado.

Radostits, O., Leslie, K. and Fetrow, J. (1994), 'Health Management of Dairy Calves and Replacement Heifers', *Herd Health – Food Animal Production and Medicine*, Second Edition, Chapter 8, p.183–227, W.B. Saunders Co., Philadelphia, US.

4
Nutrient requirements of calves

The three most essential nutrients for calf growth and development are water, energy and protein. Fibre, minerals and vitamins are also important but play a smaller role.

Water

Water is essential for all living animals and it is good husbandry to provide calves with as much fresh, clean water as they want. Weaned calves can drink 10–15 L/day and up to 25 L/day on hot summer days.

Milk contains 87–88% water, which should be sufficient for normal body requirements. Milk-fed calves will not suffer from the absence of extra water unless they are exposed to heat stress. However, as soon as they start eating solid feeds, particularly dry feeds like hay or straw, calves require continuous or regular access to fresh water. This simple practice will increase their intake of solid feeds and so reduce their age at weaning.

Overseas producers often include water in concentrate mixes to produce a slurry that allows feeds to bypass rumen digestion and so be better utilised by the young animal. Work to date has been mainly with intensively fed lambs.

To standardise the description of feed intake, it is usually expressed in terms of dry matter (DM). This is easily measured by placing feeds in an oven at 100°C for up to 48 hours. Animals eat quite similar amounts of DM no matter what the type of feed is offered. Maximum DM intakes are directly related to live weight in growing calves at the rate of 2.5–3% live weight per day; this can increase to 4 or even 5% live weight per day in high producing dairy cows. So, 100 kg weaned calves will eat about 2.5 kg DM/day, while 200 kg calves can eat up to 4.8 kg DM/day.

Energy

Energy is needed to maintain body temperature and to support normal body functions. This is known as the maintenance energy requirement. Any energy consumed that is surplus to this basic need is available for growth or the laying down of muscle and fat, which is called live weight gain. Maintenance energy requirements increase with live weight.

Energy requirements and the available energy in feeds are measured in units called joules, and more commonly in kilojoules or megajoules, replacing the once familiar calorie (1 calorie = 4.184 joules).

Only a portion of the gross energy in feeds becomes available to the calf following digestion. Undigested energy is lost in the faeces, while a small portion of the digested energy is lost through rumen fermentation and also in the urine.

The remaining useful productive energy is called the metabolisable energy or ME. This is the conventional measure of the energy requirements of the calf (in MJ of ME/day) and also the energy content of different feeds (in MJ of ME/kg dry matter or MJ/kg DM).

Because milk is a high quality feed that is digested efficiently in the abomasum, its energy value to the calf is considerably greater than that of solid feeds digested in the rumen. More than 90% of the gross energy in milk ends up as ME compared to only 50–60% of the gross energy in hay and concentrates.

Figure 4.1 Calves should be separated from their dams very soon after birth

The energy requirements for growth increase with age and weight but also vary with the energy content of the feed. High energy feeds, such as milk and concentrates, are used more efficiently for growth than are low energy feeds, such as medium quality pasture or hay. Because the maintenance ME requirement is constant for a particular live weight, the faster an animal grows, the higher the proportion of the total ME intake available for growth, and therefore the more efficiently the feed is used by the calf.

For example, an 80 kg ruminant calf growing at 0.5 kg/day requires 44 MJ for each kg of weight gain, and will only use 32% of its ME intake for growth. If growing at 1 kg/day, it requires only 31 MJ for each kg of weight gain and will use 52% of its ME intake for growth. Furthermore, it will take twice as long to achieve the same target live weight gain. Therefore, the slower growing calf will require considerably more ME than is apparent from the difference between these two ME requirements for growth.

Table 4.1 Requirements of weaned calves for metabolisable energy (ME), rumen degradable protein (RDP) and undegradable dietary protein (UDP) at different live weights and for different growth rates (prepared by Webster 1984)

		Live weight (kg)		
		80	140	200
Maximum DM intake (kg/day)		2.4	3.6	4.8
ME requirements (MJ/day)				
Maintenance (M)		15.0	23.0	30.0
M + 0.25 kg/day gain		18.0	27.0	36.0
M + 0.5 kg/day gain		22.0	32.0	42.0
M + 0.75 kg/day gain		26.0	38.0	48.0
M + 1.0 kg/day gain		31.0	43.0	55.0
Minimum dietary ME content (MJ/kg DM)				
0.5 kg/day gain		9.2	8.9	8.7
1.0 kg/day gain		12.9	11.9	11.5
Crude protein requirement (g/day)				
0.5 kg/day gain	RDP	170.0	250.0	330.0
	UDP	130.0	120.0	110.0
1.0 kg/day gain	RDP	240.0	335.0	430.0
	UDP	200.0	180.0	150.0
Minimum dietary crude protein content (% DM)				
0.5 kg/day gain		12.5	10.3	9.2
1.0 kg/day gain		18.3	14.3	12.1
Optimum degradability of protein				
0.5 kg/day gain		0.56	0.68	0.75
1.0 kg/day gain		0.55	0.65	0.74

The bottom line to this calculation is the cost of the higher quality feed to achieve that improved growth rate.

Calculations on the energy requirements of milk-fed calves differ from those for calves with developed rumens. A milk-fed calf weighing 100 kg and growing at 0.5 kg/day requires a total of 21 MJ of ME/day, 4 MJ/day less than if that calf was weaned. About 20% of the gross energy in milk is retained in the bodies of 2-week-old milk-fed calves growing at 0.3 kg/day, and this can increase to 26% in 14-week-old veal calves fed milk replacer and growing at 1.2 kg/day. However, only 11% of the gross energy intake is retained in the bodies of 6-month-old calves growing at 0.6 kg/day on a diet of hay and concentrate.

In other words, solid feeds digested in the rumen are only used about half as efficiently by the growing calf as milk digested in the abomasum. However, as already mentioned, the energy in milk costs up to four times that in concentrates, making early weaning onto solid feeds considerably cheaper than milk feeding.

The reason feed energy is the most important nutrient in any diet is that the efficiency of feed energy conversion into animal product is so low. Less than a quarter of the energy in milk fed to calves is retained by the animal and this falls to only one tenth of the solid feed when the calf is weaned!

Once the animal uses energy it is all lost as body heat, while some of the other feed nutrients are recycled within the animal.

Practical ration formulation is based on the principal of selecting the most appropriate ration ingredients to meet the animal's energy needs at the lowest cost. Requirements for the other nutrients, such as protein, fibre and minerals, can then be met by adjusting the concentration of the ration ingredients so that none of these limit animal performance. Table 4.1 (previous page) shows examples of calculations on the feed energy and protein requirements for weaned calves to achieve various growth rates at different live weights. This table was first presented by Webster (1984) using standard feeding tables published in the UK by Agricultural Research Council (1980). Many countries produce similar tables of nutrient requirements for their livestock. The other important one used in Australia is published by the US National Research Council (1989). A set of Australian feeding standards published by Standing Committee of Agriculture (1990) is mainly based on the UK feeding standards.

Table 4.1 shows that the ME intakes for 1 kg/day gain are only about a third higher than those to achieve growth rates of 0.5 kg/day but about double those for maintenance. This table also presents the required ME content of any ration to achieve growth rates of 0.5 or 1 kg/day. These data will be used in Chapter 8 in examples of ration formulation.

Protein

Proteins are required by the calf to maintain biological processes on a daily basis, as well as repairing tissues and forming blood.

Proteins are also an integral part of growth, such as the laying down of muscle. Most protein synthesis takes place in other body tissues such as the liver and gut wall, which are actively concerned with processing nutrients to meet the requirements of the body. These metabolic functions include such things as synthesis of enzymes and hormones, cell division and cell repair, and so require a continuous supply of different types of protein and energy.

All proteins are made up of building blocks, which are called amino acids. There are more than 20 specific amino acids needed by livestock. Feed protein is broken down by digestion into its individual amino acids that the calf absorbs and then resynthesises for its maintenance and growth.

The precise needs for specific amino acids are well-known in non-ruminants such as pigs and poultry, but there is little information available on the amino acid requirements of calves and adult ruminants. As the milk-fed calf depends entirely on its diet to supply them, the amino acid composition of whole milk would probably match its specific requirements.

Figure 4.2 There are many ways to quickly bulk feed milk to large numbers of calves

Rumen microbes synthesise many of these amino acids in the older calf, during the formation of microbial protein, while others are provided by the undegraded protein in the diet. Although the essential role of amino acids is in forming proteins, they can be used as sources of energy when they are excess to the calf's protein requirements. In other words, if more protein is fed than required, it is used for energy in much the same way as the starch in cereal grains or other energy sources.

Because protein is more expensive to supply than energy, it is important in ration formulation to provide only what is required.

The element nitrogen (N) is an essential constituent of all proteins, present at about 16% of the DM, though varying slightly with different proteins. That is why when feeds are analysed for protein, the total N content is measured then multiplied by 6.25 (or 100/16) to give the level of crude protein, or 6.38 where milk products are concerned. This is a good measure of the capacity to provide amino acids in many feeds, but, in others, much of the crude protein is in the form of non-protein N, usually simple compounds such as urea. All feeds contain some proportion of their N as non-protein N.

Adult ruminants benefit from this non-protein N when it is broken down by the rumen microbes into ammonia, and re-used to synthesise their own microbial protein. In calves up to 6 months old, the crude protein should be mainly in the form of true protein. The nearer the composition of a feed protein is to that of the protein in calf weight gains, the more efficiently it can be used by the calf for growth since the supply of amino acids will more closely match its requirement. In other words, there will be less likelihood of any amino acids limiting calf performance or of excess amino acids being wasted as protein sources.

Animal proteins, such as fishmeal, are more valuable to calves than plant proteins because their make up of amino acids more closely matches those of the rapidly growing calf. This is called the biological value of the protein for animals.

The extent to which the true protein is broken down by microbial action depends on its vulnerability to microbial attack and the length of time it spends in the rumen.

Dietary crude protein is now described in terms of two constituents: rumen degradable protein (RDP) and undegradable dietary protein (UDP). The RDP, which includes all the non-protein N and some true protein in the diet, is broken down in the rumen and then resynthesised into microbial protein at a rate determined by the energy metabolism of the rumen microbes. Two major forms of dietary protein escape into the abomasum, UDP and RDP that has been resynthesised into microbial protein. These are often called the bypass proteins in the feed.

There is a fairly constant relationship between the amount of microbial protein produced and the availability of ME.

If there is more RDP than available energy, the excess N that is already converted into ammonia will not be recaptured by the microbes. This is absorbed through the rumen wall and converted into urea in the liver. Much of this blood urea is wasted in the urine, although some is recycled back into the rumen as salivary urea.

The best way to ensure that calves efficiently use feed protein is to supply as much as possible in the form of UDP, that is feed protein escaping rumen fermentation, which passes directly to the abomasum for acid digestion.

The requirements for amino acids depend on the rate at which the calf is producing new tissue – its growth rate. As growth rate increases, so too does its requirements for RDP and UDP (in g/day) and its total dietary protein content (as % DM). These are shown in Table 4.1 for calves growing at 0.5 and 1 kg/day.

These protein requirements were calculated for a typical Hereford x Friesian steer calf; bull calves and calves from larger European beef breeds would require an additional 10–15% more UDP. The table also converts these RDP and UDP values to more practical units, namely the minimum crude protein of the diet and the optimal degradability of the protein (calculated as the percentage of RDP in the total crude protein). Younger, lighter calves require higher dietary crude protein levels and need more of their protein as UDP (they have lower optimum protein degradabilities).

There is also evidence suggesting that the degradability of dietary protein can influence the composition of live weight gain. For example, if calves consume rations providing adequate energy and total crude protein but UDP intakes is below recommended levels, growth rate may not be reduced but more of the live weight gain would be as fatty tissue and less as muscle. This has important implications for growing dairy heifer replacements because excess fat in the developing udder can reduce the potential for that udder to produce milk in later life. Early muscle growth is important for dairy beef calves to achieve target live weights for slaughter at set criteria of carcass fatness.

The supply of UDP as against RDP in the diets of milk-fed calves is not important because liquid feeds already bypass the rumen digestion, through the oesophageal groove, and so supply the necessary UDP for growth and development.

In fact, rumen development requires rumen digestion and so a supply of RDP in solid feeds will be beneficial to milk-fed calves. As all feeds must enter the rumen in weaned calves, the type of feed protein is important. Most vegetable proteins are highly degraded while animal proteins are more protected against rumen degradation. Extra processing, for example, heating or the addition of chemicals such as formaldehyde, can reduce the degradability of feed proteins.

Increasing the supply of energy to the weaned calf will increase the amount of RDP that the rumen microbes can resynthesise into microbial protein. In other words, the balance of RDP and UDP is not constant for any type of feed but will vary depending on the other ingredients in the total ration.

Feeds with high RDP/ME balances are fresh and conserve pastures, protein meals and urea. Feeds with low RDP/ME balances include cereal grains, maize silage and cereal straws. When formulating a ration for weaned calves, it is important to balance the feeds for both protein and energy.

Fibre

As mentioned in Chapter 2, rumen development in the milk-fed calf depends on its intake of solid feeds, which contain dietary fibre. Both the abrasive nature of plant material and the microbial digestion of the fibre stimulate the development of muscles in the rumen wall and the growth of the rumen papillae. All solid feeds contain fibre but the lower the quality of the feed, the higher its fibre content and the better it is for rumen development.

Highly fibrous feeds also stimulate saliva production during chewing and rumination. The saliva provides urea and minerals, such as sodium bicarbonate, that help maintain normal rumen microbial growth and development.

Fine grinding of feeds changes the physical nature of the fibre, but not its chemical analyses, and this can reduce its effect on rumen development. Mixing roughage with concentrates to assure consumption of both feeds without separation may require the reduction of particle size to the point that the physical abrasion (or 'scratch factor') in the roughage has lost its beneficial effects. The initial introduction of solid feeds should contain from 10–20% of the DM as roughage, with the particle size maintained as large as possible.

Minerals and vitamins

The two minerals of most importance to growing calves are **calcium** (Ca) and **phosphorus** (P), as both are required for bone development. They also have other, more dynamic, functions such as in muscle function (Ca) and energy metabolism (P).

One of the earliest signs of deficiencies of these major minerals is poor growth and poor appetite. As with most mineral and vitamin deficiencies, these early signs are not very specific; the calves simply don't appear to be doing well. This can also apply to calves suffering from parasitism or infectious diseases but, once these are eliminated, calves must be provided with the deficient nutrient before they respond. If they do not respond, then something else is wrong.

Calves do not possess what is often called 'nutritional wisdom'; they have no innate ability to select feeds to satisfy any particular nutrient craving. The only possible exception is sodium – cattle can sense the presence of sodium in rock salt or in drinking water at incredibly low concentrations. Despite what producers may be told, the intake by cattle or calves of mineral blocks bears little relationship to their mineral requirements. Deficiencies of calcium and phosphorus in milk-fed calves are rare. However, they can occur after weaning if calves are fed unbalanced diets.

The other major minerals for calves are **magnesium**, **sodium** and **potassium**. Deficiencies are unlikely except as complications of diseases that lead to scouring. They should be included in electrolyte solutions given as part of the treatment for scours. Table 4.2 shows the requirements of growing calves for major minerals, as summarised by Webster (1984), both as requirements in g/day and as minimum dietary DM concentrations.

Table 4.2 Requirements of weaned calves for major minerals at different live weights and for different growth rates (prepared by Webster 1984)

Mineral	Growth rate (kg/day)	Live weight 100 kg		200 kg	
		g/day	% in DM	g/day	% in DM
Calcium	0.5	12.0	0.42	14.0	0.30
	1.0	21.0	0.75	24.0	0.50
Phosphorus	0.5	6.0	0.20	8.0	0.20
	1.0	11.0	0.40	13.0	0.30
Magnesium	0.5	3.0	0.10	4.8	0.10
	1.0	4.2	0.20	6.0	0.15

Higher levels of dietary minerals are required to achieve growth rates of 1 kg/day compared to 0.5 kg/day. Most standard feeding tables have mineral concentrations of the available feeds and from these it is possible to determine whether additional mineral premixes should be included in calf diets.

Calf rearers can generally assume that purchased milk replacers and concentrate mixes contain the correct level of minerals for normal calf development. It is rare for problems to arise through mineral deficiency or imbalance but, as higher animal performance is sought by calf rearers, certain minerals may become limiting. One example of such an imbalance could arise from the low calcium content of most cereal grains when fed in large amounts to maximise growth rates.

Selenium deficiencies have been observed in certain regions of Australia that, in severe cases, can show up as 'white muscle disease'. Although selenium requirements in the diets are very low (only 0.1 parts per million or ppm), there is a complex interrelationship between selenium and vitamin E. For example, vitamin E can be destroyed in the rumen by oils. Therefore, if cod liver oil is included as a good source of vitamins A and D, vitamin E deficiencies can be induced.

Selenium is toxic to the calf at levels not much above the maximum required so very careful mixing into the feed is necessary. With many of the minerals, care is required to ensure that the feed mixtures only contain what is required.

If calves are grown for veal, it is important to monitor dietary **iron** levels. Calves are born with low reserves of iron and it is very low in whole milk. Additional iron supplements can increase both blood haemoglobin levels and growth rates in young calves.

The colour of meat in calves is largely influenced by its iron status. White veal is very pale because the calf becomes anaemic and lacks those meat and blood pigments that are high in iron. Pink veal allows for higher intakes of iron through concentrate feeding, but they should be kept low enough to ensure pink rather than red coloured meat. White veal is normally produced on a diet made up entirely of whole milk or milk replacer, although veal producers are now being pressured to include solid feeds in white veal diets. A further discussion on iron in calf diets and a list of iron contents in Australian feeds is included in another book I have written (Moran 1990, see below).

Calves are born with very low reserves of **vitamins** A, D and E and, hence, are very dependent on colostrum to supply these vitamins. Most milk replacers have enhanced levels because of its importance to calf health. The milk-fed calf is also unable to synthesise its requirements for the complex of B vitamins and these are normally added to milk replacers. Once the calf has a functioning rumen, it is capable of supplying its own B vitamins, and these are not normally added to concentrate mixes.

References and further reading

Agricultural Research Council (1980), *The Nutrient Requirements of Farm Livestock*, No. 2, Second Edition, Agric. Res. Coun., London, England.

Ministry of Agriculture, Fisheries and Food (1984), *Energy Allowances and Feeding Systems for Ruminants*, HMSO, London, England.

Moran, J. (1990), *Growing Calves for Pink Veal. A Guide to Rearing, Feeding and Managing Calves for Pink Veal in Victoria*. Dept. Agric. Tech. Rep. 176., Melbourne.

National Research Council (1989), *Nutrient Requirements of Dairy Cattle*, Sixth Edition, Nat. Acad. Press, Washington, DC, US.

Standing Committee on Agriculture (1990), *Feeding Standards for Australian Livestock. Ruminants*, CSIRO, Melbourne.

Thickett, B., Mitchell, D. and Hallows, B. (1988), *Calf Rearing*, Farming Press, Ipswich, England.

Webster, J. (1984), *Calf Husbandry, Health and Welfare*, Granada, Sydney.

five | 5

Obtaining the calves

Dairy farmers rearing their own heifer replacements would normally only use heifer calves born to their own dairy herd. However, producers rearing bought calves, whether under contract for dairy farmers or for their own purposes, have to obtain these animals from other dairy farmers or from calf markets.

Sources of calves for purchase

It is preferable to buy calves directly from their property of origin as this avoids buying animals that have been mixed with other calves on trucks or in yards. The farm of purchase should have a high standard of calf management and hygiene. Arrangements could be made with cooperating farmers to offer a premium to ensure calves have been offered and have drunk sufficient high quality colostrum.

In the seasonal calving areas of southern Australia, calves are often sold for a set price per kilogram live weight at 'calf scales'. These operate during the calving season at most towns in the dairying areas, sometimes two or three days each week. A recent innovation for dairy farmers is a pick-up service where operators visit the farm and buy calves direct from the farmer.

It is more convenient for dairy farmers to sell their bobby calves or excess heifer calves at these pick-up services or at calf scales, rather than truck them to calf auctions held less frequently in the larger towns. These calves are destined for slaughter within a day or two and would generally be bought by the one abattoir agent. However, arrangements can be made with the calf scales operators to buy suitable animals for rearing and individual animals can often be selected from a larger group.

The choice of animals at calf auctions is better than at calf scales but their price can be higher, depending on their demand and the intensity of bidding. Because these animals come from greater distances with age ranges greater than those sold at calf scales, they may have been in contact with more disease organisms.

To ensure less management problems during rearing, it is important that calves purchased at calf scales have good levels of maternal antibodies. Some commercial calf rearers in the US even use the field test kits described in Chapter 3 as one of the criteria for selecting suitable calves.

If animals come from interstate, it is important to check with local Department of Agriculture offices for legislation regarding movement of livestock between states.

Under the National Livestock Identification Scheme (NLIS) all calves, with the exception of bobby calves for slaughter, are required to have either a Breeder or Post-breeder Tag attached to the right (off-side) ear. These tags are small button ear tags; white for Breeder and orange for Post-breeder. Each tag has a unique identification number partly made up of the Property Identification Code. Post-breeder Tags are used to permanently identify cattle that are not identified with a Breeder Tag but are no longer on the breeder's property. All cattle will continue to be identified with an approved tail tag or large ear tag printed with the consigner's Property Identification Code (tail tag number).

Selection of calves

The following checklist can assist producers when selecting suitable calves for rearing.

- **Age and weight.** Calves should have dry umbilical cords, be no less than 4 days old and preferably up to a week of age, and weigh 40–45 kg.
- **Breed type.** The most suitable breed type depends on their eventual fate. Calves may be reared for veal and slaughtered at less than 6 months of age or reared for grass or feedlot finishing and slaughtered at 1 or 2 years of age. Heifer calves may be reared specifically for vealer mothers. In other situations, the breed may have already been specified, such as purebred Friesian for dairy heifer replacements or for a specific dairy beef market.

    When selecting calves destined for veal, purebred Friesian or beef x Friesian calves are equally suitable, although purebred Friesians are generally cheaper. Purebred or crossbred beef calves are also suitable but Jersey or beef x Jersey crossbreds do not grow as well and have poorer meat yields. Our veal trials at Kyabram have shown little difference in performance in crossbred as against purebred Friesian calves, despite the crossbreds costing up to $30 more per calf. Overseas studies show that, compared to traditional dairy breeds, later maturing beef breeds perform better and produce higher yielding carcasses. At present, the most important criteria for breed selection for veal in Australia appears to be calf price.

    Purebred Friesians may not finish on pasture as easily as crossbreds. Producers should decide on the specifications of the end product before selecting a particular breed type for steer beef. For example, one particular feedlot in Victoria sometimes fattens dairy beef steers for export to Japan and specifies purebred Friesians for feedlot entry, although it also accepts Angus x Friesian steers.

If rearing calves for vealer mothers, the most popular breed types are Hereford x Friesian and Angus x Friesian, although purebred Friesian have been shown to perform well. The later maturing breeds do not make ideal vealer mothers because their heavier live weights increase overall herd feed requirements.

- **Sex.** When rearing calves for meat production, bulls grow faster and are more efficient feed converters than heifers or steers. There is a developing Friesian bull beef industry in southern Australia producing high quality beef. Bull beef is also an ongoing system in Europe, but it does require more intensive management than steer beef.
- **Coat.** The calves' coat should be shining and clean, as dull, dry coats indicate possible poor growth due to digestive disorders, particularly if calves are obviously more than a week old.
- **Skin.** It should be clean and supple as dry skin indicates malnutrition or scouring. Twist the skin on the neck of a suspect calf; if it returns slowly to its original position, the calf is dehydrated.
- **Nose and eyes.** These should be clear and damp with no discharges, which could indicate pneumonia. A dry nose may indicate an abnormal temperature or a sick calf.
- **Head and ears.** Calves should be able to turn their head easily and their ears should be 'alert'. If the horn buds are obvious, the calf is likely to be more than a week old.
- **General appearance.** The navel and joints should not be swollen. There should be no evidence of hernias in the umbilical area; an opening into the body cavity in the umbilical area could develop into a hernia. The stomach should not be distended. Calves should be capable of sucking. If rearing for beef or veal, calves should have good conformation and not be too 'leggy'.
- **Induced calves.** In seasonal calving areas, some dairy farmers artificially induce calves to reduce their spread of calving. These calves make poor milking cows and so are generally sold soon after birth. They are more susceptible to diseases because they are smaller and given less opportunity to drink colostrum from their mothers. They are often easy to identify within groups of calves and should not be purchased.
- **Other criteria to reject calves.** Do not purchase any calves resting in the corners of pens, particularly those with rapid breathing and/or high temperatures. Do not buy cull calves from known veal growers or calf rearers. Cheap calves are usually expensive to rear.

In summary, the best calves to buy are the hardest to catch!

Price and availability of calves

The price of calves greatly depends on their supply. This varies with the area and the season. In Victoria, for instance, peak calf supply is in spring in Gippsland and the northern irrigation area, while it is in winter in the western districts. In states with year-round calving in dairy herds, calves are available throughout the year.

In previous years, the auction price for calves has been very dependent on the export returns for bobby veal. A reduction in export returns in the late 1990s led to Friesian calf prices dropping to as low as $30/calf. In more recent years, demand for export bobby veal has been high, leading to Friesian calves selling for up to $150 each.

As alternative systems to bobby veal become more established to utilise this wasted resource, other factors may influence calf prices. One good example of this was the doubling or trebling of calf prices in the UK during the 1960s and 1970s in response to the increasing demand for purebred and crossbred dairy calves for intensive finishing in systems such as 'barley beef'.

Legislation regarding marketing, transport and slaughter of bobby calves

With increasing community concern about animal welfare, various state and federal agencies are producing codes of practice on accepted farming practices for the welfare of cattle. In 1992, the Federal Government published one (SCA 1992), while in 1998, the Victorian Government published another (NRE 1998). The following is the Victorian code of practice as it refers to the welfare of bobby calves.

Introduction

- A bobby calf is defined as a calf not accompanied by its dam and under the age of 4 weeks.
- The basis of good commercial management of bobby calves for veal is the proper care and attention to the health and welfare of the calves.
- Due to their size and age, bobby calves are particularly sensitive to conditions of husbandry and transport. Consideration should always be given when bobby calves are sold to ensure the shortest practical time from sale to slaughter.
- People in possession of, and handling, bobby calves have a responsibility to care for the welfare of bobby calves under their control and this care should be separate from the interests of economic production.
- The sale of bobby calves to private organisations for fund raising should be discouraged unless competent stockmanship can be demonstrated.
- Transporters should ensure that animals reach their destination as speedily as possible, within the confines of the road law, and in a condition not significantly

less than the condition they were in when assembled for loading. The possibility of either injury or illness to the animals during transport should be reduced to a minimum. Good management and skilled driving are important to the welfare of animals carried by road or other transport.

Selection and handling

- It is desirable to present bobby calves for sale that are bright, alert, strong, vigorous, able to stand on their own, capable of being transported and at least 4 days old. Bobby calves should have been fed on the farm within six hours before delivery to a sale or pick-up point.
- The minimum recommended live weight for bobby calves being sold is 23 kg at the point of sale; obviously immature, dopey and listless calves should not be presented for sale.
- Sick or injured calves are to be given appropriate treatment or be humanely destroyed. They are not to be presented for sale, transport or slaughter.
- Handling of calves should be carried out in a manner that will avoid injury or unnecessary suffering. Calves are not to be kicked, beaten, pulled, thrown, 'dumped' or prodded with any sharp instrument. The use of electrical goading devises or dogs when handling, driving, drafting, weighing, loading or unloading is not an acceptable practice.
- Calves treated with veterinary drugs and/or agricultural chemicals shall be withheld from slaughter according to the manufacturer's recommendations. Bobby calves intended for slaughter should be fed milk or milk replacer, not milk from cows treated for mastitis or other ailments. Bobby calves that require treatment for diarrhoea should in general be treated with electrolytes in preference to antibacterials.
- The umbilical cord at the junction with the skin shall be dry and shrivelled. Cords that are fresh, wet, raw, pink or 'green' indicate excessively young calves that should not be presented for sale or transport. Bobby calves that have had their cords removed and/or treated should be individually inspected by the person responsible for the calves for evidence of dryness. Drying of the umbilical cord by artificial means must not be done. Particular care needs to be taken with the welfare of calves that are born premature.

Holding facilities

- These include on-farm holding facilities, public calf sale areas, pick-up facilities (including mobile operations), calf scales and abattoirs.
- Facilities should be constructed to permit the safe loading and unloading of calves.
- Holding pens should be constructed to provide floor surfaces that are dry, sanitary, non-slip and capable of being cleaned; holding pens need to provide

shelter from wind and rain at all times.
- The handling of calves at calf-scales and calf pick-up points should be conducted humanely and efficiently.
- The operation of calf-scales and pick-up points and the transport of calves to saleyards or direct to an abattoir should be coordinated to permit slaughter of bobby calves within 30 hours of leaving the farm.
- Places where bobby calves are held (public sales, pick-up facilities, scales and abattoirs) should have facilities and/or contingency plans to feed calves in the event of delayed removal or slaughter.
- Bobby calves that are not collected from the pick-up points by 8.00 a.m. (0800 hrs) on the day following the day of offering should be fed by the person in possession or custody of the calves at that time and, thereafter, be fed at least once a day.
- In any event, calves should be fed at least every 24 hours. Fresh or stored whole milk or reconstituted milk replacer will provide all the essential nutrients; milk replacers should be reconstituted according to manufacturers recommendations.
- Milk and milk replacers should not be fed in excess of body temperature (39°C).

Figure 5.1 Properly designed loading ramps are essential to minimise injury during transport

- To minimise the transmission of disease and to have feeding utensils in hygienic condition, it may be necessary to clean the utensils for calves between feeds.
- Calves should have access to suitable drinking water.
- Bobby calves treated for ailments subsequent to leaving the farm, with drugs or other chemicals requiring a withholding period, must not be forwarded for slaughter within the prescribed withholding period.

Transportation

- All bobby calves should be fed on the farm within six hours of transportation for sale.
- The driver of the vehicle is responsible for the care and welfare of all animals during transportation, except when either an attendant appointed by the owner or an agent of the owner travels with the consignment.
- Owners or owners' representatives should not present for transport animals that are either ill, in a weakened state or injured; the driver of a transport vehicle should not permit the loading of such animals.
- Exceptions to the above recommendation are animals that are either ill, in a weakened state or injured and requiring transport either to or from a place for veterinary treatment.
- Animals that either become ill or weak, or are injured during transport should receive appropriate attention and treatment; if necessary, they should be slaughtered humanely.
- Whenever possible, bobby calves should be transported directly, by the shortest route possible, from the point of sale to the abattoir.
- The time interval from the farm to abattoir should ensure slaughter at an abattoir by the next day.
- Vehicles used for transportation of bobby calves should be thoroughly cleaned prior to loading and at the end of every journey.
- Transport operators should check calves en route at least once every three hours.
- Calves should be transported in transports with enclosed fronts.
- Bobby calves should be loaded at a density so as to allow all calves to lie down while being transported.
- Bobby calves should be transported in separate compartment from other classes of stock.

Specific responsibilities at abattoirs

- Animals that arrive either ill, in a weakened state or injured should be isolated and receive appropriate attention and treatment as soon as possible. If moribund or seriously injured, they should be destroyed immediately.

- Bobby calves are to be slaughtered on the day of delivery to the abattoir, or within 18 hours of delivery. The first kill of the day is to include calves present at the abattoir. The kill should be in order of arrival.
- Where the slaughter of calves is delayed overnight or when calves are carried over until the next day, they must be fed as soon as practicable after the delay is known, inspected at maximum 12 hourly intervals, and be killed first at the next kill.
- In the event of an industrial dispute, leading to withdrawal of labour, notice of the dispute should be presented to management two working days before labour is withdrawn. This is to ensure that all bobby calves on hand and those being transported to the abattoirs are slaughtered within the required 30 hours.
- Where there is an extended (or unknown) delay in the slaughter of calves, abattoir management shall inform buyers to stop sending calves to that abattoir, redirect any calves in transit to alternative abattoirs and inspect all calves at a maximum on 12 hourly intervals. They must also find alternative kill sites for calves onsite and calves arriving, and/or start the kill as soon as possible after it is clear that an extended delay is to occur. They must also observe the recommendations on feeding requirements, methods and intervals as detailed under the heading 'Holding facilities' (p. 45).
- Abattoirs must have on hand sufficient feeding equipment and feed (milk replacer) to feed at least 20% of the largest possible kill days. Abattoirs must have ready access to feeding equipment and feed (milk replacer) for the largest number of calves likely to be onsite for each of the two following days. Abattoirs must have sufficient pens with appropriate shelter for the largest kill expected, and access to material (e.g. straw, rice hulls) for bedding in the event of an extended delay kill.

Other guidelines for transporting bobby calves

Although not specifically referred to in the Victorian code, the following are additional instructions for transporting bobby calves:

- Yards and loading facilities should be constructed so they do not cause injury to animals. The upper level of the loading/unloading ramp should be level with the floor of the transport. There should be no space between the loading door of the transport and the end of the ramp.
- The behaviour of livestock and the convenience of the stockmen should be considered when siting and directing lights at the loading/unloading points of saleyards.
- Ramps should not be inclined at a gradient in excess of 30 degrees. Inclined

ramps should be fitted with solid sides and have a floor surface that minimises slipping in order to prevent injury during loading and unloading.
- Floors should have non-slip surfaces. Siding should be constructed so there are no protrusions or sharp edges and should constitute a readily visible barrier to animals. Gates should be fitted with recessed closing devices or chains.
- Water should be available to livestock in the holding and delivery yards. Pens should be constructed to permit adequate shelter for calves at all times and should be roofed.
- Livestock in saleyards should not be overcrowded. Facilities should be provided to allow for the isolation of sick or injured stock. Pens should be well drained.
- Calves should be able to stand and have sufficient freedom of movement to regain standing if they fall down. Overcrowding, tethering and tying of legs and transportation in the boot of cars in unacceptable.
- Packing of animals either too loosely or too tightly in trucks predisposes them to injury; partitions should be used to reduce the likelihood of injury.
- Guidelines for the loading density of week-old calves in road transports are not readily available. For the movement of large numbers of 100 kg calves, the recommended floor area is 0.34 m^2 per calf or 80 calves in a 12.2 m (40 ft) long deck of a road transport. The acceptable loading density is reduced to 76 when calves average 125 kg live weight.

Bobby calf declarations

In 2001, the Victorian Government introduced a bobby calf declaration required for the sale of all calves to ensure that all sectors of the bobby calf industry complied with the expected industry welfare and residue standards when consigned for sale and slaughter (NRE 2001). In this case, bobby calves are defined as being under 6 weeks of age and not accompanied by their dam.

It comprises two declarations:

- For the vendor, or person responsible for the husbandry of the calves, to sign off on statements relating to the welfare and residue status of calves for sale. If calves had been treated with veterinary drugs or chemicals or had access to milk from treated cows, the vendor must indicate what the cows had been treated with and if the milk withholding period had expired. To allow for tracing back of residues in slaughtered calves, the property identification code on the ear tags must also be indicated.
- For the receiver, or person taking delivery of the calves, to sign off on statements relating to their status (transporter, stock agent or purchaser) and that the calves were presented in acceptable condition, with regard to suitability for sale. This

then passes the responsibility for their welfare during transport from the vendor to the receiver.

This declaration emphasises that all sectors of the bobby calf industry have responsibility to care for the welfare of these animals and this care should be separate from their interests of economic production. Furthermore, any false or misleading statements may attract civil action by the purchaser. This declaration then provides legislating bodies with powers to take action if antibacterial residues are detected in slaughtered calves and/or if due regard was not given to calf welfare during rearing or transport.

On arrival at the rearing unit

Whether born on-farm or bought off-farm, every calf should be identified with an ear tag or ear tattoo and weighed. Calves can be injected or dosed with vitamins A, D and E as a precaution against insufficient colostrum being drunk soon after birth. They can also be dosed with a probiotic, such as solution of naturally occurring lactic acid bacteria (non-pasteurised yoghurt), to help establish a better microbial population in the gut. It may be too late to put a dab of iodine on the navel but some farmers routinely do this soon after the calf is born.

Figure 5.2 Facilities must be appropriate for efficient rearing of calves

Calves can be tested for blood Ig levels to assess their immune status (see Chapter 3). If found to be low, the best thing may be to sell them for slaughter or at least be aware of their increased susceptibility to diseases.

Newly arrived calves should not be fed for at least four hours after transportation and preferably should be left until the following day. Following the stress of transportation, a good rest is more important than a good feed. The first drink may just be an electrolyte solution. Calves should be slowly introduced to warm whole milk or milk replacer within 24 hours of arrival.

It is important to keep newly arrived calves in a dry, warm area for at least the first two–three weeks, and isolated from older animals. This protects the bought calves against diseases transmitted by animals already on-farm but it also protects resident animals against introduced diseases. Purchased calves have had the opportunity of picking up a variety of infectious agents and the stress of transport may allow these to incubate during this isolation period and so be identified before the groups are mixed.

Other welfare guidelines on calf management, such as castration, tail docking, dehorning and humane slaughter are discussed in Chapter 12.

References and further reading

Natural Resources and Environment (1998), *Code of Accepted Farming Practice for the Welfare of Cattle*, Bureau of Animal Welfare, Agnote AG 0009, Melbourne.

Natural Resources and Environment (2001), *A Guide to Completing the Bobby Calf Declaration*, Melbourne.

Standing Committee on Agriculture, Animal Health Committee (1992), *Australian Model Code of Practice for the Welfare of Animals. Cattle*, SCA Rep. Series No.39, CSIRO, Melbourne.

six 6

Milk feeding of calves

The diversity of climate, calving seasons and milk returns in Australia has created a wide range in calf rearing systems. The length of time that calves remain housed, the method and level of milk feeding, the type of solid feeds offered and the age and weight at weaning, therefore, vary widely.

The simplest rearing system involves putting young calves out to pasture, giving them access to trees or a simple shed for shelter and feeding them whole milk to appetite (ad lib) from a trough or feeding drum for up to 12 weeks of age, but with no additional concentrates. Such a system appears to work extremely well during warm, dry weather with calves grazing top quality, spring pastures. In adverse weather conditions, or if pasture quality is sub-optimal, it can lead to older age at weaning and poor early post-weaning growth. Although labour and capital costs would be low, feed costs are high, even if the milk was valued at only 20c/L. Calf losses and disease costs may be acceptably low when the system operates effectively but could be very high if it breaks down.

The other extreme would be to house calves for the first three months and feed them limited milk (or milk replacer), specially formulated concentrate mixes plus low quality roughage. This encourages early rumen development and also achieves high pre-weaning growth rates. Following early weaning at 5–6 weeks of age, depending on concentrate intake, calves are still housed to allow greater control over nutrient intake. Once given access to pasture, concentrates would be fed for several months to minimise stresses arising from the change in their basal diet to grazed pasture. Such a system maximises post-weaning growth rates and is the basis for well-grown dairy heifers. Labour and capital costs are high but feed costs low or, at least, equivalent to the outside/pasture only system. Calf losses and disease costs should be acceptably low provided the system operates effectively.

In both cases, it is the people who rear the calves, not the system. The calf rearer teaches the animals to drink and decides on when and how the milk should be fed. This chapter describes many such systems, and their advantages and disadvantages. It concentrates on practical issues, but these depend on the principles of calf growth and nutrition more fully described in previous chapters.

Teaching calves to drink

If calves are to be fed from buckets or troughs they will have to be taught to drink. Because of their natural inclination, calves will learn to drink from teats more easily.

A calf can be trained to drink from a bucket by backing it into a corner, standing astride its neck and placing two fingers, moistened with milk, into its mouth. As the calf starts to suck on the fingers, gently lower its mouth into the bucket of milk, taking care not to immerse the nostrils so it will not inhale the milk. Keep the palm of the hand away from its nose and as the calf starts to suck the milk, gently withdraw the fingers. Hold the bucket or have it supported about 30 cm from the ground.

This process should be repeated until the calf is drinking by itself or until it has drunk at least half a litre of milk. You may need to help the calf for several feeds. It is easier to train calves using warm milk, changing to cool milk when they are drinking satisfactorily.

When training calves to drink from a teat, it should be attached to a tube that is filled with milk. As the calf starts to suck, lower the tube into the bucket of milk. The calf is usually able to keep the supply by suction. It is easier for calves to learn to feed from self-closing teats, since milk remains in the tube between bouts of sucking.

Great care must be taken not to over feed calves, especially in the first few weeks of life, as scours will result. Calves are often fed according to their live weight; 10% of their weight per day as fresh milk is the accepted amount.

The choice of liquid feeds

Colostrum and transition milk

The term colostrum is generally used to describe all milk not accepted by milk factories. However, a more correct term for milk produced after the second milking post calving is transition milk. This milk no longer contains enough Ig to provide maximum immunity to calves, but still contains other components, which reduce its suitability for milk processing.

Milk from newly calved cows should not be put into the bulk milk vat for up to eight days after calving. Regulations vary between states and between different situations. For example, some milk companies in Victoria recommend milk should not be sold for human consumption for four days after normal calving and for eight days after induced calving. During this period, cows will produce considerably more colostrum or transition milk than that consumed fresh by her calf. If only rearing heifer replacements, the colostrum produced by cows that have calved down bull and cull heifer calves would then be available for milk feeding.

Using a 25% heifer replacement rate and 45 L colostrum and transition milk per cow available for heifer rearing, this can provide up to 180 L milk available per reared calf. These calculations take into account any milk used for early feeding of bull and cull heifer calves. There should be little need for dairy farmers to buy milk replacer or use marketable whole milk to rear their heifer replacements. Increasing numbers of farmers are saving considerable money through modifying their transition milk storage systems to minimise the need to feed marketable milk or milk replacer to their heifer calves. (See Chapter 14 for one case study.)

However, some dairy farmers in Australia still only feed colostrum and transition milk during calf rearing then discard it after a few days, rather than preserve it for later feeding. Since it has little economic value, colostrum and transition milk could be readily obtained from these farmers.

This milk has the greatest value when fed fresh or within a day or two from milking. It can be stored in a refrigerator for a week or so, or in a freezer for up to 12 months. In most farm situations, neither method is very practical for routine storage, except for a small supply of frozen colostrum for emergency use with newborn calves.

There is little difference in the Ig levels in frozen compared to fresh colostrum. Only the first colostrum produced immediately after calving from older cows should be frozen for later use as a source of Ig. The ideal method to freeze the colostrum is in 1 L plastic bags placed in flat trays. This will produce wafers of colostrum about 2–3 cm thick, which can be rapidly thawed in lukewarm water. Used and cleaned 2 L plastic milk containers are also convenient. Very hot water should not be used to thaw the frozen colostrum because it can reduce its effectiveness in providing Ig.

Extremely bloody colostrum or colostrum from cows freshly treated for mastitis should not be stored, although it can be fed fresh – the latter to calves, not to be sold.

Natural fermentation is an excellent way for storing transition milk for feeding as a source of cheap nutrients. It must be handled in clean containers to prevent contamination and should be kept in plastic or plastic-lined containers with lids. Old stainless-steel milk vats are also ideal. If stored below 20°C, the natural fermentation will make the milk acid, stopping spoilage for up to 12 weeks. In warm conditions, preservatives may need to be added. These include propionic acid or formaldehyde. The stored milk should be stirred every day to maintain uniform consistency and fresh milk should be cooled before adding to it. The preserved liquid will develop a characteristic odour, but calves will continue to drink it provided they are not abruptly switched from fresh milk or milk replacer to stored milk. They may refuse to drink it if it becomes too acidic. In this case, its palatability can be improved by neutralising it with sodium bicarbonate or baking soda at the rate of 10 g/L milk.

Fresh colostrum has slightly greater feed value than whole milk so less can be fed or some warm water can be added to feed at the same rate as whole milk. When teaching calves to drink stored transition milk, it may be easier to begin feeding it diluted with warm water (hot water will curdle it) and then gradually change to cool, stored milk when calves are drinking more confidently. Calves will continue to drink such stored milk long after the rearer can't bear to get too close to it.

When the supply of stored transition milk begins to run out, fresh milk or milk replacer should gradually replace it over a week or so to give the calves time to accept their new diet. Changing from fresh milk or milk replacer back to stored transition milk can reduce intakes and lower growth rates.

Whole milk

Whole milk is the ideal food for calves. It has a high energy value and the correct balance of protein, minerals and vitamins for good calf growth and development. Health problems are generally lower when feeding whole milk compared to milk replacer as there is guaranteed quality control of the sources of protein and energy and there is no need to have to follow recipes to ensure the correct strength for proper feeding.

Calves fed whole milk are less prone to scours than those fed milk replacer. Although it is common practice to feed mastitic whole milk to heifer replacement calves, recent evidence suggests that this could lead to an increased incidence in herd levels of mastitis in later years.

Whole milk and milk replacer can both be preserved by acidification for easier feeding management. Formalin can added at the rate of 1–5 mL/L milk or hydrogen peroxide at the rate of 5 mL/L milk. Acidification can be achieved through adding 1.5 g citric acid/L milk or including a buttermilk culture (or non-pasteurised yoghurt) to ferment the milk. If the milk is made too acid, the calves' daily intake will be reduced.

Milk replacer

To many producers, the decision of whether to feed whole milk or milk replacer during rearing depends largely on its cost. In regions where the bulk of the milk is used for manufacturing purposes, dairy farmers generally feed whole milk to their calves. In such seasonal calving areas, farmers may feed milk replacer to calves born out of season because it could then be the cheaper alternative. When farmers mainly supply the liquid milk market, milk replacer is commonly used year-round.

The consistent quality of the milk replacer fed and its convenience are often other factors influencing its use. Some farmers are concerned with the marked variation in milk replacer quality from batch to batch. Even though whole milk may be cheaper, it may not always be readily available for feeding to calves. For example, the calf feeding area may be some distance from the dairy. The composition of calf milk replacers and their feeding value relative to whole milk will be discussed in more detail in Chapter 7.

The choice of feeding methods

Young calves should be run together in small groups of 6–10 calves to easily identify animals requiring extra assistance when drinking. Some producers like to individually crate (0.6 x 1.0 m per crate) or tether their calves for the first two weeks to ensure all animals are drinking and to provide quicker diagnosis of disease or poor performance. This also prevents the spread of disease between animals and, more importantly, between older and younger calves. Individual crating or tethering will reduce the incidence of pizzle (or ear, navel and udder) sucking, which often occurs in very young calves run together in groups.

Running calves into individual stalls just for bucket feeding eliminates any problems of fast drinking calves poaching milk from other buckets. The use of self-closing yokes is an alternative method. Small or timid calves should be given the same opportunity to drink similar volumes of milk as bigger or more aggressive animals.

There are a variety of systems used for feeding whole milk or milk replacer. All will produce good calf growth and weaning weights if followed correctly. The major

Figure 6.1 Feeding time in the calf shed

difference between any two systems is usually the result of the calf rearer rather than the system. Calves can drink from individual buckets or from communal troughs with or without rubber teats. Some advantages and disadvantages of these systems are:

- **Buckets** remove competition between calves for drinking space. By using one bucket per calf, each animal can receive a measured volume of milk, thus ensuring even milk intakes and calf growth rates. Small calves and timid drinkers can be given preferential treatment. However, labour requirements are higher and it is more time consuming than communal troughs.

    To ensure the oesophageal groove will function properly and direct the milk to the abomasum, place the base of the bucket at least 30 cm above where the calf is standing.

- **Troughs** allow for feeding anywhere on the farm and not just in calf sheds. However, there is less control over individual milk intakes as calves drink at different rates and more aggressive calves have the advantage. Calves should be started on buckets, then confined to a small yard to feed for a few days until they get used to trough feeding. Groups of calves will have more uniform growth rates when matched for drinking speed and age or size. Each animal should be allocated a feeding space of 35 cm or, if using rubber teats, one teat per calf.

    One innovative calf rearer in northern Victoria has modified 45 cm diameter metal pipes into a series of troughs to ensure calves can only drink the same quantity of milk. He uses individual feeding stalls to allow only one calf per 35 cm space. Metal partitions have been welded into the pipe, limiting the volume of milk available to 4 L/calf (for once daily feeding). The trough rotates, so that when the milk was poured into it, it fills each compartment very quickly. He then rotates the trough upwards into the feeding position with a handle so the calves can drink their share of milk in whatever time it takes them. He has four feeders, allowing him to feed 80 calves in just four minutes. The troughs are easily cleaned with water and then rotated to empty and dry out.

- **Rubber teats** give no additional nutritional benefit over bucket feeding as the speed of drinking milk has little effect on its utilisation. However, the production of saliva is greater in teat-fed calves and it may help maintain fluid intake in scouring calves. Teat feeding has also been shown to reduce the incidence of pizzle sucking in calves housed in groups. More capital is required in setting up the system and more labour is required for feeding and cleaning. Farmers often needlessly replace worn teats, but as long as the teat can be kept clean, it does not matter if the end has been chewed off.

    Farmers often prefer using teats into buckets because of the ease with which calves will learn to drink from teats. To many calf rearers, it seems illogical to provide both a teat and a bucket for each calf during milk rearing, as it doubles the cost of feeding equipment, greatly increases the time calves take to drink their allocated milk, then requires more labour to clean the equipment after use. Furthermore, it is easier for faster drinking calves to poach milk from their pen-mates simply by pushing their mouth away from a teat than pushing their head out of a bucket. Calves can consume 4 L milk from a bucket in less than 30 seconds compared to more than one or two minutes if using teats.

One way of group feeding calves using teats is with a suckle bar. This can be made from 50 mm PVC piping fitted with milk line entries and self-closing teats. Milk is poured into one end and sucked out by the calves. It saves carting milk and is easy to wash.

- **Calfeterias and feeding drums** are used with rubber teats and can feed large numbers of calves quickly. Because the milk can always remain covered, they can be fed away from shelter. The calf controls the amount of milk taken per feed so scouring is usually reduced provided the total milk supply is consistent. They can then be used for ad lib feeding.

With calfeterias, the teats are either positioned in a metal frame that is attached to the top of the milk reservoir with plastic tubes to draw milk from inside the reservoir, or the milk reservoir allows the milk to run into the teats by gravity. Modern calfeterias are made from moulded plastic to provide a reservoir of 2 or 4 L per teat. The teats in feeding drums are positioned around the top of the drum, while the plastic tubes nearly reach the bottom of the drum. The residual milk that cannot be sucked up into the tubes is usually left to naturally ferment.

Provided the milk is regularly stirred, the feeding drum only requires cleaning out once or twice each week. Even if it becomes excessively thick, cutting the ends of the teats will allow the calves to continue to suck up the milk. The milk must be stirred every day to ensure that it does not separate into a watery layer at the bottom with most of the protein and fat floating on the top. The tubes must also be regularly checked for blockages and build-up of milk deposits.

It is preferable to provide one teat per calf, although one teat for every two–three calves can be used with ad lib milk feeding. It is important to group calves on age and size to reduce competition if providing fewer teats than calves. Careful watching of calves at feeding will soon identify the dominant animals and whether there are sufficient teats available. Poor 'doers' can be moved back to a lighter group of calves to improve their competitive ability. Groups should not exceed 20 calves (if using a 200 L drum) and the age range should be no more than 3 weeks.

Some calves, particularly younger ones, may lose interest and stop sucking before they get any milk. This problem can be overcome with self-closing teats or by providing a pressure head of milk behind the teats; for example, the drum could be mounted on a stand and some teats positioned part way down the drum.

- **Automatic calf feeding machines** are commercially available for feeding whole milk or milk replacer. They provide for ad lib feeding according to the manufacturer's instructions. It is difficult to control individual intakes so a range of growth rates would result. Newer types are designed to release milk according to individual calf electronic ear tags, thus providing consistent intakes and, hence, growth rates. Most machines have the capacity for several teats so it is possible to group calves according to drinking speed and age. Although they are labour saving, such machines are very expensive and require power and water connections.

  Their use is very limited when feeding calves for early weaning where the maximum daily intake per calf should only be 4 L whole milk or 500 g milk replacer. When compared to bucket rearing, calves can use up to twice as much milk powder for the same live weight gain.

How much milk to feed

Rumen development, by manipulation of intakes of liquid and dry feeds, to the stage where calves can make efficient use of pasture, is one of the aims of calf rearing. The quantity of milk fed and the rearing system adopted should take this into account, whilst maintaining a balance between acceptable growth, cash cost and labour input.

As discussed in Chapter 2, the more milk fed to calves, the less solid feed and the slower the rate of rumen development. Because milk is a high quality feed, the more milk drunk, the faster the growth rate. However, the efficiency of converting this milk to live weight declines as intakes increase. When fed ad lib, 6-week-old Friesian calves can drink up to 12 L/day and Jerseys up to 9 L/day of whole milk. By the time the calves reach 6 months of age, any live weight advantage in calves previously fed ad lib milk, compared to restricted milk, is lost.

With access to concentrates and good quality pasture, together with once or twice daily feeding of 4–5 L whole milk/day, Friesians should reach a suitable weaning weight (70 kg) in nine weeks and Jerseys (60 kg) in 10–12 weeks. Many farmers still use live weight as their major criterion for weaning, often feeding more milk than is really necessary.

Although ad lib **milk feeding** is more expensive than other rearing systems, this system is often justified through faster growth rates and lower labour requirements, if using drum feeding. Earlier weaning compensates for the greater milk intake of ad lib fed calves, and advocates of this system argue that it uses only slightly more milk over the whole period than does one involving restricted milk. Provided that there are no setbacks to growth, weaning can occur at 6 weeks of age. Some farmers claim to be able to wean such young calves directly onto pasture, but it is unlikely that rumen development would be sufficient and a severe growth check would be likely. If considering such a rearing system, calves would have to be fed 0.5–1 kg/day of concentrates at least until they are 10 weeks old.

A modified ad lib feeding system still used by Victorian dairy farmers involves ad lib feeding for the first 3–4 weeks of age, then restricting milk intake to 6–7 L/day until weaning at 8–9 weeks of age. By limiting milk after 3–4 weeks of age then providing concentrates and/or good quality pasture, better rumen development should reduce any growth check immediately after weaning. If using feeding drums, more teats will be necessary in the second period.

Donohue (1986) reported growth rates in calves of 0.90 kg/day on ad lib milk for the first 3 weeks of age and then 0.78 kg/day when restricted to 6 L/day from weeks 3–8. This compared to 0.5 kg/day for calves fed 5 L/day through to weaning. Modified ad lib feeding uses less milk than a total ad lib system but adds two to three weeks onto the milk feeding period.

As the cost of whole milk or milk replacer rises, there is increasing pressure for **low milk feeding systems** that still maintain good growth rates to achieve early weaning. This is possible by feeding only 4 L milk/day, ad lib concentrates from the first week of age together with low quality roughage. When calves are eating 0.5 kg/day concentrates, milk feeding ceases. This can occur between 3–6 weeks of age. As with any system, milk feeding can be stopped abruptly or reduced steadily over the last week.

The quality and the palatability of the concentrate is the most important single factor in this system. It should be coarsely ground or pelleted. Inclusion of molasses or a sweetening agent can improve its palatability. To encourage early consumption, a handful of the concentrate should be placed in the bucket as the calf finishes drinking.

Ideally calves should be individually constrained at feeding because every calf should only drink 4 L milk every day and increase concentrate intake to about 0.5 kg/day before it can be weaned. If the calves are group fed such that the dominant calves consume more than their allocation of milk, they will appear more developed but their rumen will be smaller and they will eat less concentrates than other smaller calves only drinking their milk allocation. Therefore, group feed intakes are not a reliable indication of individual intakes.

After weaning, consumption of concentrates should increase to 2 kg/day until the animals are 3–4 months of age. Concentrates can then be gradually withdrawn, provided good pasture is plentiful. This early weaning system is low cost and has minimal labour requirements once milk feeding ceases.

The total amount of whole milk fed during rearing can vary from more than 550 L milk with ad lib feeding with no concentrates down to only 150 L plus 80 kg concentrates with restricted once daily feeding. The implications of the threefold range in milk intakes at various milk prices will be discussed in Chapter 14.

Other aspects of artificial rearing

Milk temperature

The most natural way to feed calves is to teat-feed the milk at 39°C twice daily. Milk temperature is not important provided it is consistent from day to day. It is easier to train young calves to drink warm milk and then change to cool milk. However, very cold milk removed directly from the vat should be warmed using hot water prior to feeding. If this is done, ensure that the calves are still offered the same quantity of whole milk.

Feeding frequency

Twice daily feeding is still the normal routine on many dairy farms, but once daily feeding commencing within the first week of life is adequate. Calves grow equally fast on either frequency when fed the same level of milk each day. Since the competition for milk may be stronger with once daily milk feeding, it is essential that each calf gets its fair share of milk. Correct grouping of calves is very important if feeding from a communal trough, as is at least one teat per calf if using rubber teats.

It is important to provide access to concentrates after the first week and to ensure it is fresh each day. Clean water should also be on offer, as calves will drink more water than when fed twice daily. It is possible to strengthen milk replacer mixtures to ensure smaller calves still consume enough nutrients when fed the larger volumes once daily.

Calves fed only once each day will eat more concentrates at an early age since they have more time to get hungry and seek out other feed. Furthermore, calves can be fed at the most convenient time of the day rather than after morning and afternoon milking, as is necessary when feeding twice daily. Once daily feeding should not reduce the frequency with which calves are inspected. When calves are more than 2 weeks old, it is even possible to feed them six days each week, giving the rearer a day off each weekend.

Milk dilution

Farmers often dilute milk, either to warm it or as part of a treatment for scours. Dilution of milk or milk replacer reduces the intake of nutrients due to the calves' limited gut capacity. Whether it is useful in treating scours is also questionable.

Some farmers even dilute milk when weaning calves so the animals will have the same volume but less milk solids. Calves can, in fact, be abruptly weaned off milk with no serious after-effects.

Antibacterial residues

It is essential that calves sold for slaughter do not show any antibacterial (or antibiotic) residues. Bobby calves destined for slaughter at a week of age should not be fed milk from cows treated with antibiotics unless the required withholding period for each chemical is strictly observed.

The withholding period is the time following treatment, during which products derived from any treated animal should not be used in food production. This varies for particular drugs with the route of administration into the cow (injection, oral, intramammary) and the dose rate. For most antibacterials, the withholding period for sale of milk is considerably shorter than that for sale of meat, which can be up to 30 days from administration. To be on the safe side, consider 30 days as the minimum withholding period for calves fed milk from cows given intramammary drug treatment.

Antibacterial compounds get into calves from three main sources:

- Sick calves that have been treated, usually for scours.
- Calves suckling cows that have been treated with intramammary preparations or by injection.
- Calves consuming antibiotics through suckling cows that still contain 'dry cow therapy' preparations at calving, usually due to failure to massage the preparations into the udder when initially administered, or if the cow has only had a short dry period.

Calves that are intended to be reared as replacement heifers but fail to thrive are often sold along with other bobby calves. These calves are a particularly high risk group for antibacterial residues, as they will often have been treated for some illness.

Calves are often sent to slaughter within days of being treated for scours with antibiotics or sulphonamides. In many cases, treatment with an antibacterial drug may not be necessary. Electrolytes, glucose and fluid replacement are the important components of an effective treatment for 'white scours' in calves. Antibiotics and sulphonamides should only be used on the advice of veterinarians, and withholding periods are as long as 28 days for some sulphonamide calf scour tablets.

To protect Australia's bobby veal export markets, it is vital that the withholding periods for all drugs used are strictly observed.

Mastitic milk

Milk from cows after antibiotic treatment for mastitis or other bacterial disease cannot be sold and must be discarded. Estimates in the US are that this amounts to 22–62 L/cow. Feeding this milk to calves is one way to capture some economic value from an otherwise wasted resource. This milk is often called blue milk, because of the blue dye used to identify, hence, separate it from vat milk.

Controversy still exists as to whether feeding this milk to replacement heifer calves increases their likelihood to mastitis in later life. If calves are individually penned, there is no evidence of increased mastitis. The antibiotics do not adversely affect milk digestion, increase the likelihood of greater antibiotic resistance in future disease outbreaks nor have any long-term detrimental effects on production or health. There is conflicting evidence on the potential of mastitis bacteria to increase the incidence of future mastitis in group-fed calves that can suck the developing mammary glands of other heifer calves. For this reason, farmers may wish to discard it or feed it to male calves.

Mastitic milk should not be fed to sale calves without due regard to the withholding period of the antibiotic. Calves should not be fed milk from cows caused by E. coli or pasteurella unless it has been pasteurised.

Concerns about viable pathogens in waste milk have led to large dairies in the US installing pasteurisation plants to treat all whole milk destined for calves. An assessment of the costs and returns indicates that such plants would need to be used to feed 300–400 calves before becoming economically feasible. Heating milk to 65°C for 30 minutes is sufficient to kill all pathogenic bacteria.

Labour

One of the major factors influencing the choice of feeding method is its labour requirement. The time taken in milk feeding and washing can vary from 30 seconds to three minutes per calf per day, and even longer in inefficiently run systems. One of the quickest systems involves ad lib feeding of naturally fermented whole milk from a series of feeding drums or troughs for large numbers of calves run together in a paddock or pen. In contrast, twice daily bucket feeding of milk replacer for small groups of calves in a shed is one of the slowest.

In seasonal calving regions, farmers aim to reduce the spread of calving to facilitate mating management and will even induce late calving cows. A six–eight week calving period would be the ideal for spring calving Victorian herds. This means that all replacement heifers are reared within a three–four month period – more than 200 calves in large dairy herds. Calf rearing generally coincides with hay or silage making, as well as mating and early lactation feeding. Clearly, an efficient calf system is most desirable, with less emphasis on minimising milk intakes and more on daily time management.

If feeding time can be reduced by one minute per calf on a farm rearing 60 calves, that amounts to one hour less labour each day of rearing. Remember that reducing milk feeding to five–six weeks rather than the more usual 8–10 weeks also considerably reduces the total rearing time per calf. In my 1994 survey of calf rearing practices on 320 dairy farms in northern Victoria, I found farmers spent 2.4 minutes/calf/day, or 165 minutes/calf for the entire 9.8-week milk-feeding period (Moran 1995). This can be compared with 3.9 minutes/calf/day, or 109 minutes/calf, reported in a previous South Australian study in which calves were weaned at 4 weeks of age (Wickes and others 1972).

This scenario contrasts with year-round calving herds in which calves are continually being reared but in much smaller numbers at any one time. More time can be spent on calf management, and the facilities for feeding and housing, being much smaller, can be more sophisticated. As milk can return more per litre in these regions, feeding systems would be directed more towards minimising intakes of milk or milk replacer and less towards reducing feeding time per day.

Pizzle sucking

The problem of pizzle sucking is common amongst artificially reared calves. Young calves are instinctively curious and as well as drinking, eating and ruminating, they use their mouths for all sorts of apparently abnormal behaviour such as licking and chewing inedible (or unswallowable) objects, sucking the ears, navels, teats, tails and pizzles of neighbours and even drinking urine.

Cross sucking is a potentially dangerous way of spreading infection, while compulsive urine drinkers tend to show abnormalities of rumen development. In intensive rearing systems such as group-fed veal production, these calves invariably show slow growth and poor feed efficiency.

The incidence of calves sucking each other can be reduced by providing greater opportunity for them to satisfy this desire such as feeding with rubber teats rather than buckets and using ad lib feeding drums or automatic milk feeders to give calves continual access to milk.

Hanging a piece of chain in the pen of group-housed calves may also be effective. Another, more drastic, measure would be to attach a weaning device, or metal ring with spikes on it, to its nose. Pizzle sucking calves can be individually penned or tethered during milk feeding, then offered concentrates immediately after they have finished their milk allocation.

Trying out a new system

Whenever farmers visit other farms, they generally look at how things are done there and maybe consider changing their practices to include any potential improvements they have seen. This is fine as long as they can be confident it *will* improve productivity and profitability on their farm, or maybe even make life easier.

When considering changing some aspect of calf rearing, rearers have an ideal opportunity to closely compare the 'old' with the 'new'. They should be encouraged to change practices in just one or two pens and see how the calves perform in comparison with their existing system. But they must make sure they are comparing 'apples with apples' not 'apples with oranges'. For example, if changing to an early weaning system, it would seem logical not to compare calves at different weaning ages, but at the same age, when their rumens are fully functioning. Using live weight at 12 weeks, or even older, is the best way to compare different milk feeding practices.

Multiple suckling using dairy cows

The cheapest way of feeding whole milk to calves is to allow them to harvest it themselves by suckling cows. The ratio of suckler or nurse cows to calves should be adjusted so that each calf receives at least 4 L/day of milk. The milk production of the nurse cows should then be checked to ensure adequate milk supply for her calves.

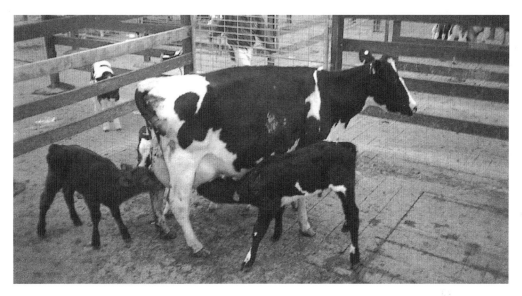

Figure 6.2 Multiple suckling reduces feed costs

Once one batch of calves is weaned off the cows, another batch can be multiple suckled. Growth rates of suckled calves are as good or even better than those achieved with artificial rearing, but they can be more variable since there is less control over individual calf intakes. There is a serious risk of infecting calves with certain diseases carried by cows. Multiple suckling should not be used for rearing heifer replacements on dairy farms where Johne's disease has been identified or where there is a high threat of the disease. Coccidia and salmonella organisms can also be transferred to calves through multiple suckling.

Cows with active mastitis infection should not be suckled because calves can transfer the mastitis, causing organisms to spread to other teat quarters and also to other cows. However, mastitic cows destined for sale could be used to foster bull calves for meat provided the cows were isolated from other cows.

Nurse cows and heifers generally produce more milk while heifers reach peak milk yield quicker when suckled than when machine milked.

Research has shown that cows foster rearing two or three calves for the first 8–12 weeks of lactation often produce more milk when returned to the milking herd than cows run in the herd from calving. Nurse cows also seem easier to break into the shed routine after a short period of suckling.

There are two types of systems for multiple suckling, continuous or foster suckling and restricted or race suckling.

Continuous suckling

This involves fostering extra calves with the cow's own calf. All calves should be matched for age, size and vigour. A proportion of cows will not adopt other calves, and such calves will steal milk from more cooperative cows. This will reduce their milk supply to their own foster calves and can lead to variable growth rates in both groups of calves. In one instance, a nurse cow rejected foster calves and, probably due to an increase in milk supply, her own calf died from scours.

For continuous suckling to work, each cow and her adopted calves must become bonded as a family unit so that the nurse cow will accept all her own foster calves but still reject others. Once this bonding has been established, it is difficult to introduce a new calf into the family to, say, replace one that died.

To help develop this bonding, the cow and calves should be kept together in a small paddock for about 10 days, and the cow should be restrained in a race or bail daily for three days at feeding to make sure all her calves have been accepted. It is sufficient to starve the foster calves for about 24 hours and then constrain the cow, unmilked for about 12 hours, with the calves for an hour each time. Other mothering systems involve keeping the calves in small pens then locking the cow in with her calves for an hour or so every day for the first week.

It can help if the nurse cow becomes confused after calving about which is her own calf. It can be removed and replaced with other calves that have been smeared with a strong smelling substance, such as neatsfoot oil, which has been placed on the cow's muzzle and also on her own calf. Some farmers also use baler twine or swivel chain and collars to tether one or two foster calves to the cow's own calf for a few days; in this case, the calves should be no more than 30 cm apart.

To maximise growth and rumen development, the calves can be given access to quality feed (grass and/or supplement) by creep grazing using electric fences. When weaning some calves early, they must be the adopted calves as the nurse cow could reject them if her own calf was removed first.

Foster suckling has the advantage that the cows can be run away from the dairy, leaving closer paddocks for the milking herd. It also allows the continued use of good breeding cows past their prime as milkers, low-testing cows or cows that do not fit the daily routine (for example, because of temperament or milking speed). Such cows have been known to milk for 18 months and rear a dozen or more calves. However, their calves tend to become wild because of lack of regular human contact and they may be difficult to train for milking. Nurse cows are less likely to cycle and this increases the spread of calving in seasonal calving herds.

Apart from any initial fostering problems, which can be time consuming, it requires minimum labour input.

Foster suckling is used more for rearing beef calves than for dairy replacements. Moss and others (1978) in Queensland used nurse cows to rear four Friesian crossbred calves for 10 weeks, after which two were weaned; the remaining two calves were fostered for a further 32 weeks. The suckled calves weighed 65–70 kg at weaning at 70 days and 275 kg at sale at 300 days. Long-term suckling should be restricted to cull dairy cows because of anoestrus problems and, hence, poor fertility.

Restricted suckling

The second system involves separating the calves from the cows, except at milking time when they are brought together in a small yard or in a suckle race. Up to four calves can suckle any one constrained nurse cow. A suckle race will restrict movement of cows better than a yard and, hence, allow smaller calves better access to available teats. The race can be made of 50 mm galvanised pipe construction, 75 cm wide, with one rail each side 76 cm off the ground. Moveable pipe barriers can be inserted into the race every 1.8 m to separate cows. The floor should be concreted for at least 1 m outside both sides of the race to prevent the ground from becoming boggy. A suitable race is shown in Figure 6.3.

To minimise teat damage, suckling should be limited to 15–20 minutes per session and cows should only be suckled for three–four weeks at a time. All quarters of each nurse cow should be suckled dry. Scours can be more of a problem with suckled calves because of the increased likelihood of overfeeding. Hygiene problems are all eliminated because the milk is harvested directly from the cow. It is important to group calves on age and size to reduce competition. With very high yielding cows and large numbers of calves to rear, it is possible to divide the calves into two groups and feed each group only once each day.

There may be little saving in labour compared to artificial rearing because calves have to be brought from the paddock or calf shed to the milking parlour each time. Cows have to be selected, such as mastitic and freshly calved cows, and then drafted from the rest of the herd. Some cows are difficult to train to accept calves, such as those that continually kick. Others are better suited for restricted suckle rearing than for

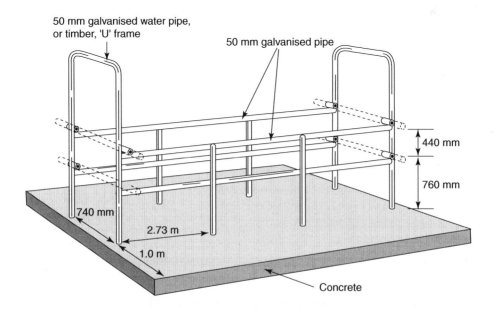

Figure 6.3. Design features of a suckling race

machine milking, such as cows with three functional teats, poor udders or slow milkers. Cows can also be rotated between the dairy and the suckling race and still run together in the milking herd. Because nurse cows produce more milk, they could lose more weight in early lactation and, hence, may require better feeding than those being machine milked.

Variations to this system are to allow the cow's own calf to suck her dry after each machine milking for the first week after calving. Alternatively, cows and heifers can be race suckled each afternoon by fewer calves and then machine milked each morning. These variations prevent milk accumulation in the cow's udder, which can have detrimental effect on yields later in lactation, while rearing several calves. The improvement in milk yield after these calves are weaned generally compensates for the milk previously taken by the calves.

A commercial producer in South Australia reared up to 180 beef calves per year using a 12-cow multiple suckling system. Cows were bailed up twice each day, with three calves allowed to suckle each cow in the morning and another two in the afternoon. After three months, the calves were weaned and another batch introduced. Three lots of three-monthly batches each year gave a nine-month lactation, followed by a three-month dry period.

Nurse cows do not begin to show oestrus after calving as soon as cows that are machine milked. To maintain a 12-month calving interval, calves should be removed from the cow for 24 hours about eight weeks after calving. Cows will normally show signs of oestrus within the next seven days and can be mated at this or the next oestrus 21 days later.

Early weaning requires strict rationing of milk so it may be difficult to combine this with multiple suckling. However, this can be done successfully by 5 weeks of age by gradually reducing either the time of access to the cows or the number of nurse cows. Calves should be weaned onto good quality pasture together with 1–2 kg/day of concentrates. The protein content of the available pasture should determine whether the concentrate is boosted with additional protein or is basically an energy supplement.

References and further reading

Davis, C. and Drackley, J. (1998), *The Development, Nutrition and Management of the Young Calf*, Iowa State University Press, Ames.

Donohue, G. (1986), *Calf Rearing*, A series of five Agnotes, Vic. Dep. Agric., Melbourne.

Donohue, G., Stewart, J. and Hill, J. (1984), *Calf Rearing Systems*, Vic. Dep. Agric. Tech. Rep. 96, Melbourne.

Green, J. and Master, A. (1990), *Calf Rearing and How to Go About It*, Northern Herd Devel. Co-Op, Cohuna, Vic.

Moran, J. (1995), 'Rearing Dairy Heifer Replacements – How Have Systems Changed in

Northern Victoria?' *Proc. Aust. Large Herds Conf.*, Albury, NSW, p.187–94.

Moss, R., Orr, W. and Stokoe, J. (1978), 'Multiple Suckling for Rearing Dairy Replacements and Dairy Beef', *Proc. Aust. Soc. Anim. Prod.*, 12, 222.

Moss, R. and O'Grady, P. (1978), 'Effect of Multiple Suckling on Live Weight, Milk Production and Fertility of Dairy Cows', *Proc. Aust. Soc. Anim. Prod.*, 12, 224.

New South Wales Department of Agriculture (1984), *Raising Dairy Calves*, NSW Dep. Agric. Agfact A1.2.2, Sydney.

New Zealand Beef Council (1990), *Producing Better Beef of Dairy Origin*, NZ Beef Coun., Auckland.

Tasmanian Department of Primary Industries (1991), *Rearing Dairy Replacements. A Manual for Dairy Farmers*, Dep. Prim. Ind., Hobart.

Wickes, R., White, B., Lewis, D. and Radcliffe, J. (1972), 'Rearing Once-Daily Fed Calves Using Differing Milk Fat Percentages, Feeding Methods and Weaning Ages', *Proc. Aust. Soc. Anim. Prod.* 9, 303–8.

Wishart, L. (1983), *The Dairy Calf in Queensland*, Qld. Dep. Prim. Ind., Brisbane.

seven 7

Calf milk replacers

About 8000 tonnes of milk replacer are produced every year in Australia. Assuming an average of 20 kg of milk replacer is required to rear each calf, this corresponds to about 400,000 calves per year reared on these milk substitutes. As this constitutes 50% or more of the total number of artificially reared calves, milk replacers is an integral part of Australian calf rearing systems.

The composition of milk replacers

A good quality milk replacer should be similar in composition to whole milk. It should contain the nutrients that calves can digest and in the right proportions. Most milk replacers should form a clot in the abomasum and so provide a slow release of nutrients to the duodenum.

In Australia, milk replacers are generally formulated from by-products of dairy processing, together with animal fats plus added vitamins and minerals. Whole milk powder consists mainly of lactose (36–40%), fat (30–40%) and milk protein (28–32%). The protein is principally made up of casein but also includes whey proteins, albumin and globulin.

The by-product of butter making is skim milk, which consists mainly of lactose and all the milk proteins; it has only half the energy value of whole milk. Whey, the by-product of cheese making, consists only of lactose, albumin and globulin, and is even lower in nutritive value. When used as the basis of milk replacers, additional fats are required.

Commercial milk replacers usually contain 20–24% protein. Young calves can only digest proteins of milk origin such as those from skim milk and buttermilk powders. The degree of processing of these powders affects the calves' ability to digest this protein. Excessive heating denatures the protein, leading to poor clotting in the abomasum and rapid passage of milk into the duodenum. Spray dried milk powders, manufactured at lower temperatures than roller dried milk powders, are the preferred source of powder for milk replacers.

In the past it has been possible to test whole milk and skim milk-based milk replacers for their clotting ability using rennet, which can be obtained from cheese factories or as junket tablets from supermarkets. Powders that have not been excessively heated will form curds, whereas those overheated will not. The curd test is as follows:

- Dilute the rennet to 1% concentration (1 mL rennet added to 99 mL water).
- Reconstitute 60 g milk replacer powder in 500 mL warm water (at 35°C).
- Add 5 mL diluted rennet to the powder solution and stand at 35°C in a bowl or sink of warm water.
- Curds should form within 30 minutes.
- This can be compared with 500 mL of warm, whole milk to which rennet has been added.

A good curd will set like jelly and maintain the shape of the container when tipped. A partial curd will be soft, like yoghurt, and will not maintain its shape when heaped on a spoon. A solution with no curd will stay liquid. Whole milk or skim milk-based replacers (those based on casein) that do not form curds should not be used in calf rearing. Several modern day milk replacers are now based on whey protein concentrates, rather than milk powder. Whey proteins do not clot in the abomasum and are digested in the intestines. Therefore, their clotting ability gives little guide to their nutritive value.

Milk replacers usually should contain 15–20% fat – the type of added fat used will influence its utilisation by the calves. Tallow (a by-product of abattoirs) is the most common fat to include in Australia because vegetable oils, which contain high levels of polyunsaturated fats, can cause scouring in young calves. Tallow is preferred because it has a similar fatty acid composition to milk fat and is cheap. Tallow is one of the few animal by-products that can now be fed to ruminants. The fat must be incorporated carefully so that the powder dissolves easily in water and the fat globules become sufficiently small so that they do not separate out in the solution following mixing. Lecithin is usually included to assist with the incorporation of added fats and to improve their utilisation in milk replacer powders.

High quality milk replacers have a fibre content of less than 0.1%. Fibre originates from plant material commonly used to increase protein levels in milk replacers. For every 0.1% increase in fibre content in replacers, about 10% of the total protein has been derived from plant rather than milk sources.

A typical milk replacer contains 70–80% milk solids, 17–20% animal and vegetable fats (for example tallow), 2% lecithin, traces of minerals (copper, zinc, manganese, cobalt, iron and iodine) and vitamins (A, D, B_{12}, K and E) with added antibiotics or antibacterial drugs.

The inclusion of antibiotics in milk replacers is a matter of concern, particularly to producers rearing their own calves born on-farm.

New diseases can be introduced through bought-in animals, for example, a different type of scour-causing bacteria. This is why antibiotics are added to some replacer powders. In theory, calves should not be fed antibiotics as normal routine because the sooner any disease outbreak can be identified and diagnosed, the sooner the calves can be treated. Low level antibiotic feeding will mask a low level of disease so that by the time calves show any symptoms, more intense treatment may be required. Furthermore, regular use of antibiotics will increase the risk of cull calves being sold for slaughter with detectable levels of antibiotic residues in their carcasses, and so jeopardise Australia's bobby veal export trade.

Most cases of scours are caused by poor feeding management rather than infectious agents (see Chapter 10), so antibiotics, which will not be effective against viral or protozoal scours anyway, serve no purpose in most cases. By continually feeding antibiotics to calves, bacteria can develop resistance to them. This means that if a bacterial disease does break out, the antibiotics prescribed by the veterinarian may not be able to control the resistant bacteria.

Antibiotics are also added to milk replacers to stimulate feed intake. Because antibiotics show their greatest improvement when management and hygiene are not the best, their routine use can give a false sense of security, which is followed by a generally poor job in calf raising.

Powders based on milk by-products are expensive and attempts to reduce their costs through using alternative protein and energy sources have been largely unsuccessful. Soybean or soya flour is a vegetable protein by-product successfully fed to older animals, but it contains an antigen that inhibits protein (in this case trypsin) digestion in milk-fed calves. This anti-trypsin antigen can be destroyed by heat treatment prior to inclusion in replacer powders, but calf production trials to date are not promising.

At Kyabram, we fed calves diets in which the treated soya flour replaced some of the milk replacer, but it was unpalatable and poorly utilised (Moran and others 1988). This was because young calves could not digest vegetable proteins as efficiently as milk proteins.

Calves cannot digest starch in their diet until their rumen is functioning. As little as 2% starch in milk substitute diets will depress growth and increase scouring in very young calves. Replacer powders with high levels of starch are not suitable for such animals. The content of starch and the proportion of milk protein to total protein should be detailed on the milk replacer bag.

Long-term storage of milk replacer powders is important. They must be packaged properly to keep out air and moisture. They should be vacuum-sealed in a plastic bag then enclosed in a light-proof bag. Even with this protection, they are best used within six months of purchase. Good quality powders include an antioxidant to reduce the deterioration of fat during storage.

Describing quality of milk replacers

In the 1970s, a panel of Australian dairy specialists developed a set of standards for milk replacers to ensure their suitability for calves less than 3 weeks old. These standards were as follows:

- The powder should contain between 15–20% fat and at least 24% protein.
- An antioxidant should be added to reduce oxidation of the fat during storage.
- The fat should be homogenised so that 90% of the fat globules have a diameter of less than four microns.
- The milk should be pasteurised and dried at such a temperature that the concentration of non-casein protein in the milk is not less than 5 mg/g.
- The milk powder should contain not more than 0.1% crude fibre and the starch content should be stated.
- The proportion of milk protein of the total protein should be stated.
- The milk powder should be supplemented with 6000 IU vitamin A, 600 IU vitamin D and 10 mg vitamin E per kg (IU stands for international units that are used to measure concentrations of vitamin in feeds).
- The milk powder should contain 100 mg/kg of iron unless intended for veal production.

More recently, US organisation BAMN (Bovine Alliance on Management and Nutrition), developed a series of farmer guidelines for calf feeding. Their guideline on milk replacers (BAMN 1997) uses the following quality evaluations:

Dry powder

- **Colour.** Cream to light tan, free of lumps and foreign material. If powder is orange-brown in colour and has a burned or caramelised smell, the product has undergone Mallard browning (non-enzymatic browning) as a result of excessive heat during storage. If the product has 'browned', there will be some loss of nutrient quality and product palatability.
- **Odour.** Powder should have a bland to pleasant odour. If odour is characterised as smelling like paint, grass, clay or petrol, the fat portion of the product may be rancid.

Reconstituted liquid
- **Mixing.** The product should go into solution easily. Milk replacer should be mixed until all powder is in solution or suspension without clumps of undissolved powder on the surface of the solution or at the bottom of the bucket. Ingredients that are in suspension but are not soluble will settle out of solution (form a sediment) if allowed to stand without agitation. This sediment layer will be more apparent as the fibre content and/or level of added minerals and/or medication increases. In some feeding situations (automatic feeders, nipple bottles, etc.), milk replacers containing significant amounts of insoluble components may not be acceptable. Care should be taken not to over mix. If agitation is continued after the product is in solution, excessive foaming can occur or the fat portion of the product may separate and form a greasy layer on the surface.
- **Colour.** Cream to light tan.
- **Odour.** Pleasant with no 'off' odours noted.
- **Flavour.** Milky with no 'off' flavours. Some milk replacers are supplemented with organic acids. These will have a 'tangy' (sweet tart) taste. This should not be confused with the 'off' lactic acid taste found in sour milk.

The best single criterion for evaluating milk replacer is calf performance. If it is poor, more detailed evaluation of management, calf health and milk replacer quality is necessary to determine the reason for the poor performance.

The nutritive value of milk replacers

The energy content of milk replacers primarily depends on their fat contents. The added fat is less digestible than milk fats so milk replacers generally contain less energy than whole milk supplying the same amount of milk solids.

Formulae are available to calculate the metabolisable energy (ME) contents of milk products and two of these are presented below for the benefit of producers wishing to calculate the energy values and energy costs of the variety of feeds used for rearing calves.

The ME content of whole milk can be calculated as follows:

$ME = [(35.9 \times F) + (19.1 \times P) + 88.8] / TS$

where ME is metabolisable energy in MJ/kg DM of whole milk
F is milk fat per cent
P is milk protein per cent
TS is total milk solids per cent

Table 7.1 lists the ME content of whole milk at various fat, protein and total solid contents. This table presents protein rather than the solids-not-fat content because dairy farmers are paid on the basis of fat and protein yields. The solids-not-fat content can be converted to protein content by assuming a constant amount of milk lactose and minerals in whole milk, as follows:

P = SNF − 5.8

where   P is milk protein per cent
              SNF is solids-not-fat per cent

Table 7.1 shows that the ME content of whole milk can vary from 20–26 MJ/kg DM depending on its composition.

Table 7.1 Metabolisable energy content (MJ/kg DM) of whole milk varying in concentrations of fat, protein and total solids

Total solids (%)	Protein (%)	Fat (%)			
		3.5	4.0	4.5	5.0
12.5	2.5	21.0	22.5	23.8	25.3
	3.0	21.7	23.2	24.6	26.0
	3.5	22.5	23.9	25.4	26.8
13.0	2.5	20.2	21.6	22.9	24.3
	3.0	20.9	22.3	23.7	25.0
	3.5	21.6	23.0	24.4	25.8

The ME content of milk replacer can be calculated as follows:

ME = (0.23 × F) + (0.06 × P) + 14.1

where   ME is metabolisable energy in MJ/kg DM
              F is fat per cent in milk replacer DM
              P is protein per cent in milk replacer DM

Table 7.2 lists the ME content of milk replacer at various fat and protein contents. To allow comparisons with other feeds, these contents are determined on a DM basis whereas the DM content of air-dry milk replacer is 96%. This table shows that the ME of commercial milk replacers can vary from 19 to 21 MJ/kg DM depending on its composition. These calculations may underestimate the contribution of lactose to the energy value of milk replacer, particularly in powders with lower than normal fat contents.

Table 7.2 Metabolisable energy content (MJ/kg DM) of milk replacers varying in concentrations of fat and protein

Protein (%)	Fat (%)			
	16.0	18.0	20.0	22.0
20.0	19.0	19.4	19.9	20.4
25.0	19.3	19.7	20.2	20.7
30.0	19.6	20.0	20.5	21.0

The nutritive value of milk replacer (of a given composition) relative to that of whole milk (of a given composition) can be easily calculated by comparing these two tables. Furthermore, these tables can be used to calculate the amount of milk replacer or whole milk required by rapidly growing young calves.

The ME requirements for calves was discussed in Chapter 4 and specifically in Table 4.1. Milk-based diets are used more efficiently for growth than solid feeds, hence, the ME requirements of milk-fed calves are slightly lower than those presented in Table 4.1. For example, 100 kg milk-fed calves growing at 0.5 kg/day each require 21 MJ/day and this is 4 MJ/day less than if they were weaned. For the same growth rate, 50 kg calves each require 15 MJ/day, while 75 kg calves each require 18 MJ/day of ME.

Assuming they are consuming negligible solid food, 50 kg calves growing at 0.5 kg/day while fed milk replacer containing 20% fat and 25% protein (or 20.2 MJ of energy/kg DM), each require 740 g DM/day or 770 g/day of air-dry powder. If drinking whole milk containing 4% fat, 3% protein and 13% total solids (or 22.3 MJ of energy/kg DM), each calf requires 670 g milk DM/day or 5.2 L/day of milk. This particular milk replacer then only supplies 91% of the ME for the same amount of DM as this particular whole milk. The daily ME requirements for 75 kg calves growing at 0.5 kg/day would be supplied by 930 g air-dry milk replacer or 6.2 L whole milk.

The relative cost of milk replacers

Producers must decide whether to feed milk replacer or whole milk to their calves. This decision is often based on the relative cost of the two feeds. This can be calculated on the basis of cost for suppling the same total solids (cents per kg DM) or cost for supplying the same feed energy (cents per MJ of ME).

The current (2002) worldwide shortage of processed dairy products has increased the price of skim milk powder, which has resulted in more expensive milk replacers. In the early 1990s, milk replacers could be bought for $40–$50/20 kg bag, whereas now they cost $60–$70/20 kg bag.

If milk replacer were available for $65/20 kg bag, it would cost $3.25/kg for air-dry powder or $3.37/kg of powder DM. If it contained 20% fat and 25% protein, it would provide 20.2 MJ of energy/kg DM, and the feed energy supplied would cost 16.7c/MJ of ME.

Let us assume that whole milk containing 4% fat, 3% protein and 13% total solids, thus providing 22.3 MJ of energy/kg DM, was the alternative liquid feed being considered. In dairy regions with whole milk payments still on a literage basis, milk replacer would be the cheaper alternative only when whole milk cost more than $3.37/kg of DM or 43.8c/L. With compositional payments, these calculations become more complex because dairy farmers must consider milk fat, protein and total milk volume. Furthermore, in Victoria milk returns vary throughout the year depending on the proportion of the state's total milk production supplying the liquid milk market. When calculating the value of whole milk on a monthly basis, the industry often uses milk fat equivalents as the basis of their calculations.

In 1999/2000 this could vary from, say, $4.20 to $7.00/kg. For milk containing 4% milk fat, this then relates to a variation of 16.8–28.0c/L on a volume basis. For whole milk containing 4% fat, 3% protein, 13% total solids and supplying 22.3 MJ of energy per kg DM, this relates to a variation of $1.29–$2.15/kg DM or 5.8–9.7c/MJ on an energy basis. Therefore, under this scenario, whole milk would always be the cheaper option in terms of both c/kg DM and c/MJ of energy.

Using milk replacers to rear calves

When planning a rearing program based on milk replacer, it is best to order a bulk supply of the replacer as it is often cheaper per bag than smaller lots. There are only one or two Australia-wide brands of milk replacer available, while there are several others with smaller distribution networks. Quality control during processing is sometimes questioned with some of the less well-known brands of replacers, particularly when milk powders become available on the market at extreme discount prices. The generalisation that 'you get what you pay for' holds for such products. For example, one particular batch may be cheap because it was subjected to excess heating during processing.

Between 55–65% of the total cash costs of replacement heifers is attributable to feed, with 95% of the total feed costs occurring post-weaning. In this context, saving $5–$10/calf on lower-cost milk replacers does not seem to be a good economic decision if its poorer quality places the calf at greater risk of nutritional ill health.

It is sometimes possible to purchase second grade whole milk powder that has not been overheated but could be lumpy, contaminated with dirt or in some other way unsuitable for use in human feeds. Samples of these cheaper milk by-products should be sent to commercial feed laboratories for evaluation or at least subjected to the rennet test described above. There are cases when cheaper milk powders become available that are not inferior in quality. In one instance, cheap product was sold from a supplier of flavoured whole milk powders because it was a transition powder between two different flavours. Another example is whole milk powder not suitable for specific export markets because of small variations in the protein and fat specifications.

It is important that calf rearers understand mixing strengths when preparing milk replacers for feeding. The mixing instructions usually refer to the quantity of powder within a given volume of reconstituted mix, not the amount of water added to the powder. For example, the instructions may be to mix 250 g powder in warm water and make up to 2 L. If 2 L of warm water were added to the 250 g powder, the volume of the final mix would be 2.25 L and the calf would have to consume more liquid for the same nutrient intake.

In this first case, making 250 g powder up to 2 L produces a solution with a strength of 1 in 8 or 12.5%, whereas adding 2 L to 250 g powder would give a strength of 1 in 9 or 11.1%. The important factor is to make sure that the correct amount of milk replacer is measured, or preferably weighed out, for the number of calves being fed.

Once weighed out into a bucket using a spring balance or kitchen scales, the powder should be placed in a calibrated container with some water already in and mixed, either mechanically or by using a hand whisk. More water is then added to give the correct final volume and temperature. It is very important that a consistent feeding temperature be used. For warm solutions, this should be around body temperature, about 36°C, but no more. Some brands of milk replacer can be mixed in cold water and this will be indicated in the instructions written on the bag.

An alternative to weighing is to use a measure, often provided by manufacturers, where one measure of powder is equal to one feed for each calf. In this case the measure must be regularly checked, because milk replacer powders can vary considerably in bulk density and errors of up to 20% can arise.

Depending on the number of calves being fed, the liquid replacer can be measured out by hand into buckets for individual calves, poured into troughs for communal feeding or into large feeding drums if using ad lib systems with teat feeding. It can be pumped into individual buckets using a petrol bowser dispenser connected to a large reservoir. With one person feeding, say, 50–100 calves, the feeding time will average about half a minute per calf.

The provision of hot water for feeding and washing up afterwards is an important practical consideration. It is more economical to have an off-peak hot water system. The temperature of cold water can vary from 4°C (in winter) to 11°C in summer (and even higher in tropical regions). Heating the water to 70°C and mixing it with tap water, roughly in the ratio of 2:1, produces a final mix of about 40°C. This can be judged by hand but ideally should be tested each time with a thermometer.

Because milk replacer contains dried milk powders and non-milk products, it behaves differently to fresh whole milk once it enters the abomasum. Curds of whole milk, being more digestible, are broken down more quickly in the abomasum, thus allowing the calf to have more frequent drinks. However, curds of milk replacer must be given more time in the abomasum for their complete digestion.

Milk replacer should be fed less frequently than whole milk. Too frequent feeding of too much milk replacer can lead to abomasal-induced milk bloat. This occurs when the newer clot envelops the old, partially digested clot of milk replacer, reducing the opportunity of gases to escape and causing distension of the abomasum. It can also lead to overfilling of the abomasum and the spilling over of unclotted replacer into the intestines, a certain cause of calf scours. Twice daily feeding of small quantities of milk replacer can successfully rear calves, but once daily feeding is likely to create fewer problems.

Examples of several milk-replacer rearing systems

Figure 7.1 Calves can be fed their milk by many systems

During the 1960s, Sydney University researchers Jane Liebholz and Roy Kellaway worked with producers to develop rearing systems based on early weaning calves fed limited amounts of milk replacers. This system became commercialised and is now widely used in Australia. The recipe has been modified in recent years for rearers to feed milk replacer either once or twice daily, and to feed different sized calves. The once daily feeding recipe is presented in Table 7.3.

Table 7.3 Daily feeding regime for an early weaning rearing system, based on once daily feeding of milk replacer, for small (<30 kg) or large (30–45 kg) calves

Days	Small calves		Large calves	
	Powder (g)	Water (L)	Powder (g)	Water (L)
1–2	200	1.5	200	2.0
3–5	250	1.5	300	2.0
6–8	300	1.5	350	2.0
9–11	350	2.0	400	2.5
12–35	400	2.0	500	2.5

The calves are individually tethered for the first week, then group housed with continuous access to water and run into individual stalls equipped with self-closing yokes for bucket-feeding of the milk replacer through rubber teats. From day 6 onwards calves are offered ad lib concentrate pellets plus clean, long straw to stimulate early rumen development. The calves are abruptly weaned off the milk replacer at 5 weeks of age but remain group housed and fed pellets plus straw for a further five weeks. They are then given access to pasture but still fed the pellets for several weeks thereafter, depending on pasture quality.

Each calf would be expected to consume 12–15 kg milk replacer together with about 100 kg pellets over the 10 week feeding period and gain about 50 kg live weight.

Another role for milk replacers in calf rearing is through boosting the concentration of whole milk. The rationale is that calves can be fed smaller volumes of whole

Figure 7.2 Rearing 5000 calves each day requires an efficient system for milk feeding

milk yet consume similar or higher intakes of energy and protein. This would be beneficial to small calves when introduced to once daily feeding. One other advantage of this 'supercharged' milk is that the smaller volume for the same nutrient intake is less likely to reduce appetite for solid feeds, thus encouraging calves to take to concentrates more readily. It is essential to provide sufficient drinking water to satisfy the greater thirst of calves when fed whole milk plus milk replacer.

Research with different concentrations of milk replacer has shown that the optimum milk DM concentration for calf growth and feed utilisation is about 15%. Since whole milk contains 12–13% total solids, in theory only 25 or 30 g powder should be added to each litre of whole milk. However, successful systems (see Table 7.3) have been developed using once daily feeding of 500 g replacer in 2.5 L water or 300 g replacer in 2 L colostrum or whole milk.

One large-scale dairy farm in Tasmania uses 'supercharged' milk to successfully rear all their dairy heifer replacements. Following birth and 12–24 hours with their dams, calves are tethered for several days until they readily drink the 2 L of transition milk plus 300 g milk replacer offered once daily. They are then group housed and introduced to concentrate pellets and good, clean straw by day 10. By day 28, the bigger calves are weaned off liquid feeds, while the smaller calves are fed milk for another four or five days.

Calves are kept inside for eight weeks; by then each is eating up to 3 kg/day of the pellets. Once introduced to pasture, calves voluntarily wean themselves off pellets when eating sufficient pasture. Calves double their birth weight by 10 weeks of age and no whole milk is required in this rearing system. Each calf consumes on average 55 L of colostrum, 9 kg milk replacer and 100 kg pellets.

References and further reading

Bovine Alliance on Management and Nutrition (1997), *A Guide to Modern Calf Milk Replacers. Types, Use and Quality*, Arlington, Virginia, US.

Kellaway, R., Grant, T. and Chudleigh, J. (1973), 'The Effect of Roughage and Buffers in the Diet of Early Weaned Calves', *Aust. J. Exp. Agric. Anim. Husb.*, 13, 225.

Liebholz, J. (1971), *The Nutrition of the Young Calf. 1. Milk Feeding*, Aust. Meat Res. Com. Review No.2, p.1.

Moran, J., Gaunt, G. and Sinclair, A. (1988), 'Growth, Carcass and Meat Quality in Veal Calves Fed Diets Based on Whole Milk or Milk Replacer', *Proc. Aust. Soc. Anim. Prod.* 17, 254.

Roy, J. (1980), *The Calf*, Fourth Edition, Butterworths, Sydney.

Thicket, B., Mitchell, D. and Hallows, B. (1988), *Calf Rearing*, Farming Press, Ipswich, England.

eight 8

Solid feeds for calves

This chapter deals with the nutritive value of solid feeds and their formulation into diets for rapidly growing, weaned calves.

Weaned calves require feeds rich in energy to promote high feed intakes and good animal performance. The energy value of feeds is measured in terms of their metabolisable energy or ME concentration. Feed energy is usually provided by carbohydrates that come essentially in two forms: material inside plant cells such as starch (in cereal grains) and sugars (in high quality pastures), and digestible material in the cell walls such as cellulose.

Calves can digest the first form of energy themselves but require the rumen microbes to digest and utilise the plant cell walls.

Calves also require feeds that are high in crude protein (CP) and can supply undegradable dietary protein (UDP) as well as rumen degradable protein (RDP) to the digestive tract. RDP originates from the feed nitrogen, both true protein and non-protein nitrogen, which is broken down in the rumen into ammonia to provide one of the basic nutrients for rumen microbes to grow and produce microbial protein. This microbial protein together with any feed protein escaping rumen digestion (the UDP) then passes into the abomasum for digestion by the calf itself. These processes are described in more detail in Chapter 4.

The nutritive value of solid feeds

The best way to describe the nutritive value of any feed for weaned calves is in terms of its DM, ME and CP contents and its protein degradability – the balance of its supply of RDP and UDP. Unfortunately, most published tables of feed composition only report the average energy and protein contents, despite the very large variations within any one type of feed.

With plant-derived feeds, for example, these can vary with the particular season in which it was grown, the soil type and the crop management during growth. The ME and CP of animal-derived feeds can vary with the type of animal and the degree to which the feed was processed.

Very few feeding tables list protein degradability as it is difficult to accurately measure and can also vary considerably. The balance of RDP and UDP depends on many factors such as the calves' DM intake, the degree of processing as well as the total dietary ME supplied to the rumen microbes. In other words, it partly depends on the other ingredients in the ration.

Protein degradability or quality is usually expressed as the percentage of RDP in the total protein. Of greater importance to young rapidly growing calves is the proportion of UDP in the total protein.

It is better to describe protein quality of a feed in terms of its undegradability or its supply of UDP to the animal. Furthermore, rather than give each feed a single value, grouping it into one of four categories would be more appropriate. These categories are as poor, moderate, good and high.

Category	Degradability	Undegradable protein
Poor	0.71–0.90	10–29%
Moderate	0.51–0.70	30–49%
Good	0.31–0.50	50–69%
High	Less than 0.31	More than 69%

Over the last 15 years, the Victorian Department of Agriculture FEEDTEST laboratory at Hamilton has analysed thousands of samples of cereal grains, roughages and other feeds as part of its commercial feed evaluation service. These data, together with others collected from Australia-wide surveys, are presented in Tables 8.1, 8.2 and 8.3. The tables include mean values for DM, ME and CP and their ranges (where there are sufficient analyses available).

Categories of supply of undegradable protein are based on results collected worldwide, because there are presently insufficient data available in Australia. Concentrations of the major minerals calcium, phosphorus and magnesium can be obtained from books on feed composition tables such as MAAF (1987) or NRC (1989).

Table 8.1 covers energy-rich feeds. Apart from oats and sorghum, the ME content of cereal grains is fairly consistent, although their protein levels can vary widely. Energy levels are high in feeds containing large amounts of fat such as whole cottonseed, rice bran and copra meal.

Table 8.1 The nutritive value of selected energy-rich feeds. See text for description of the undegradable dietary protein or UDP supply (P, poor; M, moderate)

Feed	Dry matter (%) Mean	Metabolisable energy (MJ/kg DM) Mean	Range	Crude protein (%) Mean	Range	UDP supply
Cereal grains						
Barley	90	13	12–13	11	7–15	P
Wheat	90	13	12–13	12	9–16	P
Oats	90	11	9–12	9	6–13	P
Triticale	90	13	12–13	12	8–16	P
Maize	90	14	12–16	10	7–14	M
Sorghum	90	10	7–13	11	6–15	M
Rice	90	12	–	7	6–8	P
By-product feeds						
Whole cottonseed	90	15	–	24	22–26	M
Brewers grain	25	10	–	23	21–26	M
Copra meal	90	13	–	20	18–22	M
Rice bran	90	14	–	15	14–16	M

Cereal grains are poor sources of UDP, except for maize and sorghum. The amino acid composition of their protein is also not ideal for young animals. Therefore, cereal grains require protein supplementation if they form the basis of the diets of weaned calves. Most cereal grains are low in calcium.

Table 8.2 covers protein-rich feeds. Grain legumes are multipurpose in that they are relatively cheap sources of both protein and energy, although the protein is very degradable. Urea is a non-protein nitrogen source with no energy value and 100% degradability of its nitrogen. It is mainly used to substitute for true protein sources in compound feed mixtures. There are a wide variety of protein meals available for feeding weaned calves in Australia, nine of which are listed in Table 8.2. The meal with the highest supply of undegraded protein, fish meal, also has a good make up of amino acids for young calves, but, at $900/t (tonne) or more, is very expensive.

Meat and bone meal, the other animal protein in the table, is considered to be a good protein supplement for young calves but large differences can occur in calf growth from different sources of meat and bone meal. These differences are related to the amount of bone in the meal. Calcium content is a good indicator of bone content and this is inversely related to protein content, a good indicator of the meat content.

The outbreak of mad cow disease in Europe in the 1990s has led to an Australia-wide ban on the use of feeds derived from animal products in diets for ruminants (except tallow, gelatin and milk products). Consequently, meat and bone meal can no longer be incorporated into the diets of weaned calves or growing cattle.

Table 8.2 The nutritive value of selected protein-rich feeds. See text for description of the undegradable dietary protein or UDP supply (P, poor; M, moderate; G, good; H, high)

Feed	Dry matter (%)	Metabolisable energy (MJ/kg DM)		Crude protein (%)		UDP supply
	Mean	Mean	Range	Mean	Range	
Urea	100	0	–	250	–	Nil
Grain legumes						
Lupins	90	13	12–13	32	28–36	P
Peas	90	13	12–13	24	20–27	P
Protein meals						
Fish	85	11	–	67	65–70	H
Meat and bone	85	13	–	53	46–59	G
Soybean	85	13	–	52	46–59	M
Peanut	85	14	–	45	32–53	P
Safflower	85	11	–	43	22–54	M
Cottonseed	85	12	–	42	37–45	M
Rapeseed	85	12	–	39	33–43	M
Sunflower	85	10	–	35	30–46	M
Linseed	85	12	–	34	30–38	M

The plant protein meals in the table supply only moderate levels of UDP, while peanut meal is a poor source. The amino acids supplied in the protein of oil seed meals does not match that required by young calves as well as that supplied from animal sources. Energy levels of all the protein meals are comparable to those in cereal grains.

Certain oil seed meals contain compounds toxic to young calves. The anti-trypsin agent in soybean for milk-fed calves has already been mentioned in Chapter 7. Another one of relevance to weaned calves is gossypol in cottonseed meal. If the cottonseed meal is sufficiently heated during extraction of the cottonseed oil, this compound is destroyed. However, to be on the safe side, no more than 20% cottonseed meal should be included in growing calf rations.

Table 8.3 (following page) covers the nutritive value of selected roughages. The DM and UDP supply of the conserved feeds in the table indicate values firstly for hay and for silage. During the ensiling process, some of the true protein is converted to non-protein nitrogen, reducing the level of UDP supplied to calves.

Because silage requires less time for sun curing than hay, it can be made earlier in the season when pasture quality is higher. From state-wide surveys of quality in hays made from both grass and legume-based pastures in Victoria over 10 years, the ME in hays averaged 8 MJ/kg DM (range 6–10) and protein contents averaged 11% (range 5–19) while pasture silages sampled averaged 11 MJ/kg DM of ME and 14% crude protein. Not a lot of silage is fed to young calves in Australia, probably because it is more difficult to handle than hay and requires specialist equipment. Good silage is very nearly as good as good grass, but poor silage is very unpalatable to young calves.

Table 8.3 The nutritive value of selected roughages. See text for description of the undegradable dietary protein or UDP supply (P, poor; M, moderate)

Feed	Dry matter (%) Mean	Metabolisable energy (MJ/kg DM)		Crude protein (%)		UDP supply
		Mean	Range	Mean	Range	
Cereal straws						
Wheat	90	6	5–7	3	1–4	M
Oaten	90	7	5–8	3	1–4	M
Barley	90	7	5–8	4	2–5	M
Conserved feeds (hay/silage)						
Lucerne	85/30	8	7–9	16	14–20	M/P
Grass-based	85/30	9	6–10	9	7–10	M/P
Legume-based	85/30	9	8–10	13	10–16	M/P
Maize silage	35	10	9–11	6	5–8	P
Sorghum silage	35	8	7–9	6	5–8	P
Grazed pastures						
Grass-based, immature	20	11	10–12	14	12–16	P
Grass-based, mature	35	7	5–9	6	3–8	M
Legume-based, immature	15	11	10–12	20	16–25	P
Legume-based, mature	30	8	5–10	12	10–15	M

Cereal straws are very poor sources of energy and protein and their only real benefit in calf rearing is during the milk-feeding stage as a stimulus to rumen development. Lucerne hay is an excellent feed for weaned calves as it is high in UDP and if cut early, high in crude protein. However, like grass and legume-based hay and silages, it will have poor nutritive value if managed to maximise DM yield rather than feed quality. Yet maize silage has maximum yield and quality at the same stage of maturity.

The stage of maturity can have a quite dramatic effect on the energy value of conserved pasture. Tasmanian trials found the ME of silage cut on 15 October to be 11.0 MJ/kg DM when the perennial pasture was still in its vegetative, pre-flowering stage of growth. The ME levels decreased consistently every week from then on and

Figure 8.1 A self-feeder for dispensing calf meal or pellets

finally fell to 7.4 MJ/kg DM of ME when cut on 15 December, six weeks after ear emergence. Growth rates in 250 kg steers decreased from 1.5 kg/day when fed the early cut silage to only 0.1 kg/day when fed the late cut silage. Clearly, hay made late in the season is of little benefit to young rapidly growing calves.

Many of the leafy fodder crops such as sorghum and millet make good feeds while grazed, but if left to grow for silage, they have much lower energy and protein values. Some of these can be toxic at certain stages of development and their grazing management should be discussed with local advisory officers and consultants.

The list of grazed pastures in the table is very small, but it gives some idea of the enormous range in their nutritive value. Most local Department of Agriculture offices would be good sources of information on the best pasture mix and its optimum grazing management with young calves. Unfortunately, many dairy farmers pay too little attention to the grazing of their young stock and heifer growth rates and subsequent performance suffers accordingly. This will be discussed in Chapter 13.

Pasture quality is invariably higher in immature grass and legume-based pastures than in mature ones. The level of available cell contents decreases; hence, the level of cell walls increases as pastures mature. Calves can make better use of the available energy in immature, early spring pasture than when the pasture has flowered and set seed.

From Table 4.1 in Chapter 4, it can be seen that young rapidly growing calves require diets providing at least moderate supplies of UDP. However, immature pasture can only supply poor levels of UDP. The supply of protein to the intestine can be improved by providing additional dietary energy in the form of concentrates. Therefore, not only will the calf's energy intake be enhanced through concentrate supplements, but also its intake of intestinal protein. These additional nutrients will improve growth rates in young grazing calves.

Rumen capacity in young calves does not reach mature proportions until 5–6 months of age, and unless pasture quality is high, at least 10 MJ/kg DM, feed intakes and growth rates are restricted by limited rumen capacity. With heifers weighing less than 200 kg, dietary fibre content has a greater influence on intake and growth than energy content, whereas above 200 kg, heifer performance is less affected by ration fill characteristics. Consequently, high energy supplements are usually required to maintain good growth rates in young heifers.

Even when at their best, pastures fall short of being the complete feed for calves under 6 months of age. Multiple suckled calves can balance the deficiencies in grass by frequent drinks of milk and thus sustain high growth rates. However, even with limited access to milk, growth rates of young calves fed pasture suffer from restricted nutrient intakes, particularly dietary energy. This will be discussed later in this chapter.

The nutritive value of subtropical and tropical perennial pastures is generally poorer than those in temperate areas. Higher levels of supplemental energy and protein are required by calves reared on pasture in northern NSW and Queensland than in Victoria and other southern states.

The three tables of feed quality highlight the wide variation that can occur in energy and protein levels within any one feed. When formulating rations for young calves, or for any livestock, estimates of the quality of the diet are only as good as the information available on the ingredients. It is important that accurate values of energy and protein contents of the feeds actually being fed are used in ration formulation.

There are several commercial feed evaluation laboratories operating in Australia, but one of the better equipped ones with a rapid turnaround time for most analyses is the FEEDTEST laboratory run by the Victorian Department of Natural Resources and Environment at Hamilton – phone (03) 5573 0900. It provides reply-paid plastic bags with good instructions on sampling feeds and is one of the cheapest operating in the country.

Feed intake and calf performance pre-weaning

The contribution of solid feeds to the performance of young calves fed limited milk or milk replacer can be quite large, particularly when they are weaned at very early ages. For example, in a survey of 30 calf feeding trials in the US with weaning ages averaging

32 days (ranging from 19 to 45 days), calves averaged 0.3 kg/day of concentrates (ranging from 0.1 to 0.5 kg/day) and grew at 0.3 kg/day prior to weaning (ranging from 0.1 to 0.5 kg/day).

Table 8.4 summarises data on concentrate intakes and growth in calves fed 450 g of milk replacer in 3.8 L water each day for 21 days and then half this amount until weaning on day 28. Roughages were not fed during this period. The table presents average values each week together with the range for poor to good calves.

Table 8.4 Concentrate intakes and growth rates in pre-weaned calves

Age (weeks)	Concentrate intake (kg/day)		Growth rate (kg/day)	
	Mean	Range	Mean	Range
1	0.1	0–0.1	0.1	0.1–0.2
2	0.2	0.2–0.3	0.1	0.1–0.2
3	0.5	0.4–0.7	0.5	0.4–0.7
4	1.0	0.9–1.2	0.6	0.5–0.7
Average for 4 weeks	0.5	0.4–0.6	0.3	0.3–0.4

Feed intake and calf performance throughout the rearing period

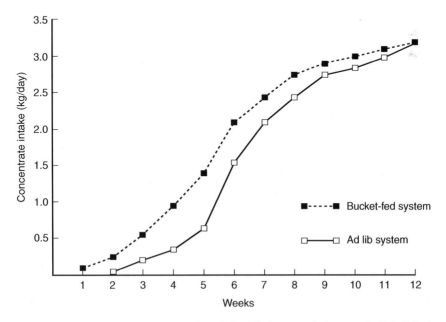

Figure 8.2 Daily intakes of concentrates when fed ad lib from week 1 to week 12 in Friesian bull calves fed either ad lib or limited milk replacer until week 5

Once calves are successfully weaned, concentrate intake rapidly increase. Rearing systems in the UK often limit access to pasture until calves are 3–4 months old. Figures, 8.2, 8.3 and 8.4 show concentrate intakes and growth rates in Friesian bull calves reared indoors on either ad lib or limited milk replacer from week 1, when they are bought at about 10 days of age. The calves fed ad lib each consumed 30 kg milk replacer; those on limited once daily feeding only consumed 12 kg.

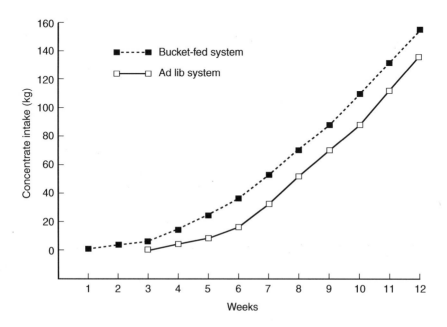

Figure 8.3. Cumulative intakes of concentrates when fed ad lib from week 1 to week 12 in Friesian bull calves fed either ad lib or limited milk replacer until week 5

Concentrate intakes increased rapidly following weaning at 5 weeks of age, particularly in calves previously fed ad lib milk replacer. By 12 weeks, both lots of calves were eating similar amounts of concentrates. Daily intakes are shown in Figure 8.2 and cumulative intakes in Figure 8.3. Each animal required 130–140 kg concentrates over the full 12-week period but this would be reduced to 100 kg or less if they were grazed by, say, 8–10 weeks of age.

Figure 8.4 shows live weights of calves in the two rearing systems. It is of interest that the extra gain made by the ad lib calves was all achieved in the first three weeks of rearing. One useful measure of quality of management of calf rearing systems is calf live weight at 12 weeks of age, which would normally mean after 11 weeks of rearing. This single measure takes into account the feeding and management during milk rearing, weaning and early post-weaning growth.

The 12 week weight in calves can vary from 85 kg or less to more than 125 kg

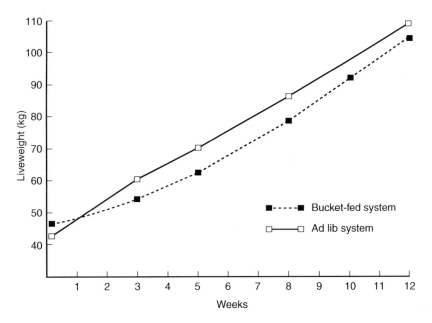

Figure 8.4 Live weights in Friesian bull calves fed ad lib or limited milk replacer until week 5 together with ad lib concentrates from week 1 to week 12

depending mainly on milk intake and the success of the transition phase from milk to solid feeds. Live weights of 95–105 kg at 12 weeks of age would be indicators of a well-managed rearing system.

Criteria for weaning calves

Different fashions in calf husbandry in different areas of the world have tended to favour different ages as optimal for weaning dairy calves, from as early as four to as late as 12 weeks after birth. However, the trend over the last 20 years has been towards weaning as early as possible. There are several good reasons for this:

- The feed energy in whole milk or milk replacer costs up to four times more than the feed energy in concentrates and up to 20 times more than in grazed pasture.
- Liquid feeding is very labour intensive and time consuming.
- Facilities for rearing calves during birth to weaning, such as pens, are more costly than those required after weaning, and the shorter the period, the fewer the pens required.
- Disease control, particularly scours, is easier to manage in weaned calves.

One of the most important practical considerations in the pre-weaning period is to ensure that the calf stays alive and that it does not succumb to disease severe enough to set back its growth. Therefore, the most economical feeding system prior to weaning may not necessarily be the one that costs the least in food and labour.

Jersey calves generally drink less milk and grow slower than Friesian calves and so have different weaning criteria. Table 8.5 presents guidelines often used by dairy advisers for weaning heifers (e.g. Donohue and others 1984).

Table 8.5 Suggested guidelines for weaning dairy heifers

Breed	Live weight (kg)	Chest girth (cm)	Age (weeks) (limited milk)	Age (weeks) (ad lib milk)
Jersey	60	87	10–12	6
Friesian	70	92	9	6

Although the rumen of a 3- to 4-week-old calf may be as effective as that of an adult animal, the rumen capacity should be the major determinant for weaning. This depends on dry feed, hence, indirectly on milk intake. For example, an aggressive calf that drinks more milk than its pen-mates will probably eat less concentrates and roughage. Therefore, at the same age, it may be heavier than its pen-mates, but its rumen will be less developed. Consequently, this particular animal should be weaned at an older age.

Calves can be successfully weaned onto dry feed when eating 0.5 kg/day of concentrates. This limit can be increased to 0.75 to 1 kg/day if producers cannot afford to have any post-weaning setbacks, such as in pink veal systems. Individual concentrate intakes are difficult to estimate in group-housed animals. However, this level of intake normally occurs around 6 weeks of age. Weaned calves should weigh at least 70 kg and be seen ruminating.

The bigger the calf when entering the rearing unit, the quicker it can be weaned. For every 10 kg increase in initial live weight, it should take seven days less to reach the same intake of concentrate.

Concentrate mixtures for early weaning calves

The first concentrate mixes offered to milk-fed calves are often called starter rations. It is becoming more common for producers to use commercially produced pellets rather than mix them from raw ingredients on-farm. This is because they must be highly palatable, fresh when fed and specifically formulated to provide the correct balance of nutrients for the transition period from milk to solid feeds.

These rations should contain at least 18% crude protein and 12 MJ/kg of ME. The inclusion of rumen buffers, such as sodium bicarbonate, in calf starters has also been shown to improve intakes and growth rates. They are usually packaged in small sized (3–5 mm) pellets. This reduces dust and ensures that calves cannot select out any ingredients they like and reject others.

Several practises are used to encourage very young calves to nibble starter rations at an early age. These include placing a small amount of pellets in each milk bucket, assuming this is the milk feeding system used, as they finish their daily allocation of milk. Calves may not initially eat more pellets but intakes can increase in later weeks by over 25% and this has been shown to improve growth rates to four weeks by 40%.

Young calves should only have fresh pellets available. They should be offered a new batch of pellets each day while those left from the previous day can be fed to older weaned calves. Twice daily feeding of concentrates is often recommended to give the slower eating calves a better chance of having their full daily ration. One large-scale dairy farmer prefers bagged rather than bulk pellets because he considers that the bulk handling equipment tends to powderise the pellets and reduce their palatability to very young calves. Fresh water must be available from the start.

One problem with milk rearing calves at pasture is encouraging them to eat starter pellets at an early age. Calves seem to prefer fresh grass (particularly in high quality spring pasture) to pellets. Feeding the starter rations in textured form, by including molasses, flaked as well as rolled cereal grains, together with some coarsely chopped highly palatable hay, may be one way of overcoming low concentrate intakes in very young calves reared at pasture.

When starter rations are combined with straw, penned calves eat more pellets because of a more stable rumen level of acidity. Research in Australia has shown that giving young calves access to straw increased their pellet intakes by 15% and their growth rates by 25%. Providing calves with better quality hay will increase roughage intake at the expense of the pellets and so reduce growth rates. The better the quality of roughage, the less pellets eaten.

Calf rearers throughout the world normally feed the straw long and not finely chopped because of its inherent 'scratch factor' (see Chapter 4). However, recent overseas research has shown that the incorporation of chopped straw (at 18% of pellet weight) into starter pellets improved both pellet intakes and calf growth rates. This would then remove the need to feed the roughage separately and would greatly reduce time and labour in feeding and cleaning pens. Other forms of roughage, such as cottonseed or oat hulls, could be used. Lupin hulls would be ideally suited because their higher digestibility, compared to other hulls, would not dilute the energy content of the starter pellets.

It must be emphasised that to be effective, the roughage must be coarsely chopped and not finely ground or milled before inclusion in the pellets. Excellent calf performance has been achieved by grinding roughage through a 22 mm screen (10 mesh) and incorporating the complete diets in 5 mm diameter pellets. The handling of chopped straw or other fibrous by-products by feed companies would require additional processing and equipment.

Commercial starter pellets cost upwards of $300 per tonne, and more if producers are in outlying areas, so it is not unreasonable to expect producers to want to mix their own rations. Twenty years ago, formulations for calf starters were complex mixtures containing milk powders, many feeds and supplements. These have since been shown to be no better than simpler grain-based mixtures supplemented with bypass protein, minerals and vitamins.

Table 8.6 Examples of Canadian calf starter rations

Ingredients	Ration 1 (%)	Ration 2 (%)	Ration 3 (%)	Ration 4 (%)
Ground hay	–	–	10	–
Rolled barley	50	–	57	50
Rolled oats	20	28	10	24
Cracked maize	–	40	–	–
Soybean meal	18	20	15	14
Lucerne meal	5	5	–	5
Molasses	5	5	5	5
Calcium phosphate	1	1	2	1
Salt	1	1	1	1
Vitamin A	2200 IU/kg in all rations			
Vitamin B	330 IU/kg in all rations			

Table 8.6 shows several calf starter rations used in Canada. These rations use soybean meal as the source of bypass protein, but this can be substituted for other oil seed meals listed in Table 8.2. If using meals with lower protein contents then levels of inclusion should be adjusted accordingly.

Rations 1 and 2 are both supply 18–20% protein and can be fed together with long straw. Ration 3 is a complete starter ration as it includes ground hay and contains only 16% protein. It can be fed without additional roughage. Ration 4 is also low in protein and can be used when feeding calves on skim milk.

Calves have been shown to adapt to urea-containing starter rations with no harmful effects. Growth may be slowed slightly, but more economical gains are possible. Additional molasses can be added to improve their palatability.

Canadian recommendations (Winter and Lachance 1983) are for these rations to be fed until 12–14 weeks of age, after which calf grower or regular high protein milking supplements can be fed.

Formulating concentrate mixes for milk-fed animals is more difficult than for weaned calves because of their requirements for liquid feeds. Provided milk feeding is based on minimal fluid intakes with weaning at 4–6 weeks of age, early weaning recipes such as those in Table 8.6 can be given. However, Australian rearing systems use a diversity of milk intakes, fed from 5–12 weeks, making it difficult to recommend specific diets that will apply in every case. Research is still required to develop suitable concentrates formulated for rearing calves at pasture, particularly for encouraging early concentre intakes.

When formulating rations for milk-fed calves, calf rearers must consider, first, that the total intake of nutrients from both liquid and solid feeds are sufficient for adequate growth rates and, second, that rumen development can occur as rapidly as the particular milk rearing system will allow. In other words, there are two separate digestive systems occurring in the one animal. In contrast, post-weaning ration formulation has only to consider digestion in a fully functional ruminant.

Formulating rations for weaned calves

Assuming the aim of calf rearing is to produce animals that can make efficient use of grazed pasture, the bulk of any post-weaning diet should be grazed pasture. In certain situations, such as pink veal, calves must be entirely hand-fed so nutrient intakes can be closely monitored and controlled. Not only is the DM intake of grazing animals very difficult to measure, but also the intake of energy, protein and bypass protein is well nigh impossible. Producers generally have to use live weight change as their best indicator of how well they are feeding their grazing calves.

Calf grower rations are used for hand feeding or supplementing weaned calves at pasture. These are lower in energy and protein levels than calf starter rations, namely 11–12 MJ/kg DM of ME and 14–16% protein. They are usually made up 80–90% cereal grains and 10% or more oil seed meal with additional salt and calcium supplements. Urea is occasionally used to partly substitute for the protein meal. Calf grower rations are quite similar in composition to the high protein concentrate pellets often fed to dairy cows. In fact, many farmers buy these pellets for feeding to both milkers and young stock.

The optimum level of protein in calf grower rations will depend on the other feeds being offered. For example, if calves are hand-fed or grazing high legume roughages, it should be possible to reduce protein levels to 12%. Whether this can be done by removing the source of undegraded protein and supplying all the crude protein as urea is often a subject for debate. Levels of crude protein and particularly UDP should not become too low in heifer diets. This can reduce the rate of muscle development in early growth and lead to excess fat deposition in the udder, and this has been shown to have long-term detrimental effects on lifetime milk production.

Although multiple suckled calves may not always be early weaned, they would benefit from additional concentrates. This can be fed out in a yard with creep fence such that cows cannot enter, or in an adjoining paddock separated by an electric fence positioned such that calves can walk under it but not the nurse cows.

The requirements for ME, RDP, UDP and the major minerals have been tabulated for calves of different live weights and growing at different rates in Tables 4.1 and 4.2 in Chapter 4. The nutrient content of different feeds has been tabulated in this chapter. The skills of ration formulation are to match these nutrient requirements with nutrient supplies from the available feeds in the most economical way. Once energy and protein demands and supplies have been matched, producers can check requirements for the major minerals calcium, phosphorus and magnesium and their availability in the selected feeds.

To formulate a mixed ration that supplies the daily amounts of RDP and UDP referred to in Chapter 4, a decision has to be made on a single value to use for the protein degradability in each ingredient. Actual degradabilities of protein were specifically not listed in the feed composition tables in this chapter because they can vary so much. However, if required, mean values of the degradability for each category of UDP supply can be used in such calculations. These would then be 0.80 for poor, 0.60 for moderate, 0.40 for good and 0.20 for high supply of UDP.

Using the Table 4.1 in Chapters 4 and the average energy and protein contents of different feeds in the tables in this chapter, grower rations have been formulated for beef crossbred bull calves weighing 140 kg in three different scenarios as follows:

1. Grazing unrestricted immature legume-based pastures, for example, white clover-dominant pastures in spring.

2. Grazing restricted mature grass-based pastures and only eating 2 kg DM/day, but with unrestricted access to a supplement of 80% rolled wheat and 20% cottonseed meal. This could occur on heavily stocked farms during summer.

3. Hand-fed a ration of 50% lucerne hay and 50% rolled barley to appetite, such as could occur during winter.

In all three cases, the calves were eating to appetite, hence, would consume 3.6 kg DM/day. Total DM intakes would be reduced if feeds were too low in quality, but this is unlikely in these three cases. Table 8.7 summarises the calculations required to predict their growth rates and the total nutrient intakes in each scenario are listed.

From Table 4.1 in Chapter 4, 140 kg calves growing at 0.5 kg/day require 32 MJ/day of ME, 370 g/day of CP, 250 g/day of RDP and 120 g/day of UDP. This should be achievable in all three scenarios. Corresponding daily intakes for 1 kg/day growth rates are 43 MJ of ME, 515 g of CP, 335 g of RDP and 180 g of UDP.

The calves grazing lush spring clover pastures would consume sufficient CP and RDP and would be only just short of ME (by 3 MJ/day) and UDP (by 36 g/day). Therefore, they would be able to grow at close to 1 kg/day. However, their DM intakes may be limited by the high moisture content of the clover, which can fall as low as 12 or 13% in very lush swards.

Table 8.7 Supplies of dry matter (DM), metabolisable energy (ME), crude protein (CP), rumen degradable protein (RDP) and undegradable dietary protein (UDP) to 140 kg crossbred calves in three scenarios. See text for further details.

Scenario	Ration ingredient	DM intake (kg/day)	ME intake (MJ/day)	CP intake (g/day)	RDP intake (g/day)	UDP intake (g/day)
1	Pasture	3.60	40	720	576	144
2	Pasture	2.00	14	120	72	48
	Wheat	1.28	17	154	123	19
	Cottonseed meal	0.32	4	134	81	54
	Total	3.60	34	408	276	121
3	Lucerne hay	1.80	14	288	173	115
	Barley	1.80	23	198	158	40
	Total	3.60	37	486	331	155

The calves grazing the mature summer pastures with supplements would be unlikely to grow faster than 0.5 or 0.6 kg/day because of low energy and protein intakes, whereas those hand-fed the lucerne/barley mix could achieve at least 0.75 kg/day growth.

This approach can be used to predict growth rates in any situation, provided the nutrient content of the ration ingredients are known or can be confidently predicted. This is particularly important when deciding on rations based on purchased ingredients as total costs of feeding calves should ideally be calculated on the basis of cents per MJ of ME or cents per g of protein.

The role for pasture with weaned calves

It has long been known that calves can be weaned onto grass alone at an early age if rapid growth rates are not desired. As already mentioned, the rumen in 3-week-old calves can digest pasture quite efficiently but performance is limited by rumen volume and, hence, voluntary intake. Large-scale surveys in New Zealand have shown that compensatory gain in poorly reared calves just does not happen. The lightest calves when weaned at 3–4 months of age were still the lightest when mated at 14–15 months of age. Early weaned calves fed concentrate at pasture may be able to partially compensate for any early growth setback but only if they have access to high quality pasture.

Young calves are more susceptible to adverse weather conditions than older calves. This can be very important if early weaning onto pasture in a wet, cool spring or during winter. The importance of adequate shelter for young calves will be discussed in Chapter 11.

Provided pre-weaning feeding practices have been directed towards early rumen development, calves weaned at 12 weeks of age onto good quality pasture can grow nearly as well as those housed and fed ad lib concentrates.

Pasture intakes are closely related to digestibility so the higher the quality of pasture, the higher the intake and the faster the growth rate. Donohue and others (1984) consider that Friesian heifers with unrestricted access to dry pastures (in Victoria), direct cut silage or poor hay will grow at 0.2–0.4 kg/day. On good hay or wilted silage they will grow at 0.5–0.7 kg/day, while on high quality, leafy pasture, growth rates can be as high as 0.7–1 kg/day. Jersey calves are smaller and eat less, therefore, they will growth at slightly lower rates. Post-weaning target growth rates for heifers will be discussed in Chapter 13.

Clearly, for optimum growth rates in grazing calves, concentrate levels and composition should be adjusted to the availability and quality of the grazed pasture and to any climatic stress encountered during rearing. The better the quality of pasture or other roughage sources, the smaller the benefits from feeding additional concentrates.

It is possible in certain situations to make blanket recommendations on concentrate feeding. For example, to achieve target live weights at mating and calving in Queensland, heifers grazing tropical pastures should be continuously supplemented with 1.5 kg molasses plus 60 g/day of phosphorus each day from 4–24 months of age, together with 0.5 kg/day grain and 0.5 kg/day lucerne hay during winter.

Calves may decide for themselves how much concentrates they require. Calves early weaned indoors but not grazed until 8 weeks of age have been found to stop eating concentrates within a week of grazing high quality, spring pasture. However, in autumn, when pasture was still of good quality but in short supply, they ate on average 1 kg/day of concentrates for the first eight weeks at pasture. Moving the pellet trough further each day from the water trough has been found to encourage calves to eat more pasture.

If pasture is short, hay can be substituted. It should be high in protein, above 13 or 14%, and be green to provide vitamin A. Good quality lucerne or other legume hay, or first quality pasture hay, is best for weaned calves. Good quality cereal hay can also be used, but additional protein and vitamin A supplements should be fed. The amount of hay fed should depend on what else they are eating and hay costs if being purchased. With poor quality pasture and cheap hay, hay can be fed ad lib. But with calves just early weaned, hay should be restricted so as not to reduce concentrate intakes. Silage is as good as hay provided it is of sufficient quality that calves will have no problems accepting it.

Calves can be set stocked separately to other stock, they can be strip grazed ahead of milking cows or they can be rotationally grazed behind the milking herd to clean up the paddocks. Many dairy farmers agist their young stock off the farm. Whatever the grazing system, it should allow for continuous growth throughout the rearing period.

There are certain animal health issues that need to be considered when grazing young stock and these will be discussed in Chapter 10.

Dairy heifers also have critical periods during their first 12 months of life when live weight gains should be restricted to 0.8 kg/day. Higher growth rates induce fatty tissue deposition in the developing udder and reduce lifetime productivity. This will be discussed in Chapter 13.

Special requirements for pink veal systems

Most commercial calf rearing pellets in Australia are not suitable for early weaning calves for pink veal production. Pink veal calves must be housed and hand-fed entirely on concentrates and roughages until slaughter at 4–5 months of age. The specialty pellets are based on ingredients low in iron to ensure pale coloured meat. They also contain relatively high levels of rumen buffers such as sodium bicarbonate, at 2–4% of the pellet weight, to stimulate appetite and growth. Several stockfeed manufacturers in southern Australia produce 'pink veal pellets' and others would on request, provided the order was sufficiently large.

Once the pink veal production system is established and good quality calves are produced, it is feasible to prepare homemade concentrate mixes directly from the raw

Figure 8.5 Group rearing of calves requires skill to ascertain the best age for weaning

ingredients and to feed it as a meal rather than pelleted. It is important to ensure that the particular management system produces acceptable meat quality before reducing total feed costs by formulating such rations. Pellets are conveniently packaged for handling but are more expensive than a meal of the same formulation. Further details about suitable ingredients for pink veal diets are provided in another book I wrote (Moran 1990, see References and further reading).

Concentrate mixes can be made up of whole cereal grains together with a pellet containing the protein, vitamin and mineral supplements. Canadian pink veal growers often feed whole maize grain plus a 36% protein pellet in ratios varying from 3:1 at 6 weeks of age through to 6:1 at 19 weeks of age (on a fresh weight basis). They found that feeding the maize cracked or rolled reduced feed intake and growth rate.

Canadian feeding trials compared whole maize, rolled barley and rolled oats grain for finishing veal calves to either 140 or 230 kg live weight (Drevjany 1986). Calves fed maize had the lowest grain requirements and the best feed conversion but required the most protein supplement. Calves fed maize required 3.2 kg DM/kg gain, which was 23% less than calves fed barley (3.9 kg DM/kg gain) and 40% less than calves fed oats (4.4 kg DM/kg gain).

Table 8.8 presents the performance of pink veal calves in Canada when weaned off milk replacer at 3–5 weeks of age and fed maize grain plus protein/minerals/vitamins. Live weights and concentrate intakes are those at the beginning and end of each four-week period, while growth rates and feed conversion ratios are the average for the entire four weeks. These particular calves had birth weights of 50 kg, 10 kg higher than most calves used for pink veal in Australia.

Table 8.8 Animal performance in Canadian pink veal systems

Age (week)	Live weight (kg)	Concentrate intake (kg/day)	Growth rate (kg/day)	Feed conversion ratio
0–4	50–58	0.1–0.5	0.3	1.0
4–8	58–79	0.5–2.5	0.7	2.5
8–12	79–108	2.5–3.2	1.0	2.8
12–16	108–141	3.2–4.2	1.2	3.2
16–20	141–175	4.2–5.5	1.2	4.1

Rumen digestion of concentrate mixes can be improved through using various feed additives, while animal performance can also be improved with growth promotants. Kyabram feeding trials have shown little benefit from commercial feed flavours in stimulating appetites of veal calves, although molasses and sugar can to be effective. The decision on which feed additives or growth promotants to use should be based on discussions with government advisers and meat processors handling the finished carcass. Withholding periods for their use should be strictly followed to eliminate chemical residues from the pink veal.

Some producers of pink veal prefer to feed milk plus concentrates right through until slaughter. This will increase growth rates and also the proportion of live weight gain in the carcass – that is the dressing percentage in the slaughtered animal. Exactly how much milk is fed depends on total feed costs, age at slaughter and, of most importance, the additional returns from the heavier carcasses.

One useful measure of performance in pink veal systems is the age at 70 kg carcass weight. This can vary from 140 days for early weaning systems to 105 days when calves are fed a maximum of 5 L/day of whole milk (consuming a total of 570 L milk). By feeding 10 L/day of whole milk, calves can achieve 70 kg carcass weight by 105 days, and in the process drink a total of 725 L milk. Although veal calves can drink up to 25 L whole milk each day, it is desirable to restrict milk intake to 10 L/day, in which case calves will eat up to 1 kg/day of concentrate.

References and further reading

Drevjany, L. (1986), *Towards Success in Heavy Calf Production*, Min. Agric. Food, Ontario, Canada.

Donohue, G., Stewart, J. and Hill, J. (1984), *Calf Rearing Systems*, Vic. Dep. Agric., Melbourne.

Ministry of Agriculture, Fisheries and Food (1987), *Feed Composition. UK Tables of Feed Composition and Nutritive Value for Ruminants*, Chalcombe Publications, Marlow, England.

Moran, J. (1990), *Growing Calves For Pink Veal. A Guide to Rearing, Feeding and Managing Calves for Pink Veal in Victoria*, Vic. Dep. Agric. Tech. Rep. 176, Melbourne.

National Research Council (1989), *Nutrient Requirements of Dairy Cattle*, Sixth Edition., National Academy Press, Washington, DC, US.

Winter, K. and Lachance, B. (1983), *Management and Feeding of Young Dairy Animals*, Canada Dep. Agric., Ottawa, Canada.

nine 9

Communicating with the calf

Success or failure in raising calves depends to a great extent on the rearer's attitude to the calves and his or her ability to react promptly to the calves' numerous signals. Interpreting these signals is a skill that can be easily learnt.

Recent developments in calf rearing are directed towards reducing the average time spent with each calf. In many cases, at least part of the time saved would be well spent in observing the calves more closely.

Signals to watch for from the calf

Don't let your senses idle when handling calves as many potential or actual problems may be picked up by close attention.

A calf getting pneumonia will have laboured breathing; one with scours will expel manure across a wide area. You can smell the white scours when you enter a calf shed or see the greenish manure of calves suffering from salmonella. A sweet, acetone-like odour will indicate that a calf, which has been scouring for a few days, has reached a stage where it is breaking down its meagre supplies of body fat in an effort to cover its energy needs.

Trained hands will identify the calf with a high temperature by touching the calf's ear, or will test the temperature of milk offered to the young animal to make sure it is neither too hot nor too cool.

A good nose will help locate a calf with a hoof infection, or a bale of mouldy hay or lumpy calf starter pellets.

Quietly talk, hum or even sing while you work to help the calves become familiar with your voice. Scratch each calf behind its ear or underneath its throat. Later on, this rapport could help you convince a sick animal to eat or otherwise cooperate with you.

Forming a bond with your calves is valuable in the long run.

When moving groups of calves around, remember not to try and move them from behind. Calves can see almost 360 degrees around them, but cannot see directly behind them. So always drive animals from the side. If you are directly behind them, they are just as likely to stop and turn around to see who is behind them. They always go in the direction they are headed, so calves should face the direction they are to be driven before any pressure is put on them to move.

When trying to get them through a gate, stand beside the gate. Once the calves are looking at you and facing the gate, step toward them and they will run through it to escape from you. Walking with the calves slows them down, whereas walking against the direction speeds them up.

Paying attention to the signals calves are constantly giving you enables you to improve your communication with them. Possibly in time, you may develop your own dictionary of 'calf-ish' language. Though it may not have too many entries, it may be of value to you from time to time.

A Canadian dairy researcher and avid 'calf watcher', Dr Lumir Drevjany, has listed over 60 entries in his particular dictionary. The same signal can be interpreted in different ways by different knowledgeable calf raisers.

Figure 9.1 Skilled calf rearing requires empathy with infant animals

With his permission, I have selected calf signals of more relevance to calf rearers in Australia. In many of these, I have also used his expressive and often quaint phrases.

There are veterinary treatments for most diseases whose symptoms become apparent through changes in calf behaviour and/or appearance, such as those described below. This chapter just lists symptoms that may indicate the cause of the stress, while Chapter 10 deals with some of the veterinary treatments. In some cases, mention is made of Dr Drevjany's suggested remedies. These include injections of antibiotics or vitamins or even drenches with unusual solutions (such as ginger or aspirins). Before embarking on such treatments, it may be best to seek additional opinions from veterinarians, preferably those with some proven expertise in calf management and disease. There are many suggested treatments for problems during calf rearing that don't always appear in the standard veterinary texts, such as using charcoal or corn flour to reduce the incidence of scouring, or using ginger as a 'tonic' for sick animals. Some of these are based on accepted medical principles, such as reducing the rate of movement of gut contents in scouring calves through increasing its viscosity. Others have evolved over the years from 'folk medicine' and as yet are either not fully understood or may even be discounted by mainstream veterinary science. That is not to say they don't work.There is no harm in discussing them with veterinarians and experienced calf rearers. In fact, this should be encouraged in the interests of better calf husbandry.

Changes in normal calf behaviour symptomatic of stress

The calf is charging your knees, running around the pen
These signs characterise a healthy calf. It will do well for you.

It has a poor appetite at birth
A disinterest in food shortly after birth is often related to traumatic events preceding or surrounding the birth. Don't wait for the problem to correct itself. Offer the calf high quality colostrum by stomach tubing two or three times a day. Continue this treatment until the calf is ready to eat on its own.

To prevent recurrence, review the nutritional program for the milking herd, particularly during the 60-day, non-lactating period preceding parturition, and correct for any deficiencies in protein, minerals and vitamins.

It is resting in an abnormal position
About one hour after feeding, walk through the shed and observe the calves. Healthy calves rest in a curled-up position with feet tucked under and heads back along the body. They appear relaxed with regular breathing rhythms. Any deviation from this standard should be judged with suspicion, although some healthy calves just rest flat on their side.

A calf that lies flat on its side may need propping up to prevent the fluid from the stomach draining back to the oesophagus and then into the lungs. If its neck is stretched directly ahead, with front feet tucked squarely under its chest and its shoulders humped quite high, salmonella may be a problem. A calf that curls its back up and down, with the nose pulled close to the body, may have a sore throat.

It does not stretch when standing up after a rest

Following a lengthy rest, calves will generally stretch their legs when aroused and get up. If it does not, pay particular attention to it, such as ensuring it drinks with its pen-mates. A lack of stretching is often the first sign of ill health.

It is disinterested in the food and surroundings

This animal could be telling you that you have betrayed it in the past and you didn't react to its previous signals. The road to recovery would not be easy. Try to remove the original source of stress and treat the infection with a broad-spectrum antibiotic. Give the calf electrolytes to prevent dehydration, using stomach tubing if necessary. If a digestive disorder is the cause, a teaspoon of ginger (apparently a known Canadian tonic) may restore the calf's interest. Inject the calf with vitamin B and check that vitamins A, D and E have previously been given.

It lies with its neck stretched, front feet tucked squarely under its chest and shoulders hunched high

This calf is likely to be suffering from salmonella. The temperature could have risen to 41°C and the calf would be generally weak and depressed. Foul smelling diarrhoea, often green in colour, contains blood and later, pieces of intestinal lining. Unless the disease is identified and treated very early, 60% of infected calves usually die and those that survive will perform poorly.

Don't introduce new calves into the shed until all animals are sold and the premises is disinfected. To prevent infection in humans, high standards of personal hygiene must be maintained. This is difficult on seasonal calving farms where alternative calf rearing sheds may not be available, but on year-round calving farms, a temporary smaller shed could be constructed.

It gulps the milk and chokes on it

This occurs in calves that are underfed, under stress or have to compete for milk. Some calves will plunge their heads into milk buckets, splashing it all over the floor and inhaling some into their lungs. Offer a small quantity of milk at a time and, if possible, separate the calf from others. The gulping and choking usually stops once a regular feeding program is established.

It stops eating starter pellets

Partial or complete refusal of calf starter may indicate that energy needs have been fully satisfied by liquid feeds. Stress or a severe case of digestive or respiratory disorders may have the same consequences. Prolonged treatment with certain drugs (such as sulphur drugs), particularly oral, may impair rumen microbial activity and, thus, temporarily reduce starter intakes.

If all calves refuse it, feed quality may be the problem. Check for mouldy and/or musty concentrates ingredients (if using an on-farm mix). Excess minerals can also reduce its palatability, which can be improved with molasses.

It is kicking the belly area with its hind legs

This indicates pain in the abdominal area. The source of the pain could be twisted stomach, constipation, urinary calculi (kidney stones) or bloat. Desperately seeking relief, the calf with a twisted abomasum frequently lies down and jumps up. To help, place the calf on its back on heavy straw bedding and, holding the front and hind legs, roll it from side to side a few times.

A constipated calf frequently strains and bellows loudly while trying unsuccessfully to pass manure. If the calf is not drinking, give it water or electrolytes. Urinary calculi could be suspect if substantial deposits of mineral salts are deposited on the sheath (of bull calves) and the calf tries to frequently urinate.

It is unable to stand or even raise its head

Examine the calf thoroughly for possible soreness, such an injured knee, displaced joint, infected navel, etc. If it cannot even raise its head, this may indicate complete exhaustion due to a long battle with pneumonia or scours. If the calf has a normal body temperature and a history of good health, it could be due to muscular dystrophy through a deficiency of selenium. Once treated with selenium and vitamin E, the calf could be back on its feet within 24 hours.

It drinks the urine from other calves

Pizzle sucking is a vice usually related to an unsatisfied sucking instinct, particularly in early-weaned calves. This problem can be largely overcome by tethering calves during milk feeding, using rubber teats or keeping calves separate till well past weaning. Hanging a piece of chain in pens may also be effective in group housing systems. Another, more drastic, measure would be to attach a 'weaning device', or metal ring with spikes on it, to its nose.

It is resting in the corner of the pen, with its head turned away from its pen-mates

This animal should not be ignored. First, get the calf up and if it stretches, it is OK. If it doesn't, then it requires attention. The calf may be at the bottom of the 'pecking order' and should be moved in with smaller or less aggressive calves. If the shed offers poor protection against the wind, this corner may be the warmest area of the pen and all calves will tend to congregate there.

It is shivering with its hair standing up along its back

This animal is suffering from cold stress and should be better protected from draughts or provided with thick, dry bedding and a source of heat. If only one or two calves show these symptoms, check their body temperatures. Calves may shiver in winter if fed milk at too low a temperature.

It shows increased breathing at normal air temperature

Increases in respiration rate in hot weather are expected. Some of the best gaining calves can have higher than normal rates as they consume more feed and require extra oxygen for its assimilation into body weight gain. Normal breathing rates are 56/minute at 4 days, 50/minute at 14 days and 37/minute at 35 days of age. In the majority of cases, calves with increased breathing have reduced lung capacity due to respiratory problems such as pneumonia. Increased body temperatures often accompany these disorders.

It is standing with its front legs spread out and head stretched ahead

These are important signs of a lengthy bout of pneumonia. Only a portion of its lungs is functional and the spreading of the legs allows the calf to try and secure more volume for the lungs to make breathing easier. If the above symptoms are accompanied by a heavy discharge from the nose and frothy saliva is running from the mouth, then damage of lung tissue has probably been irreversible. Another symptom of pneumonia may be an arched back with the calf moaning.

It grinds its teeth

You are dealing with a calf that has lost the will to live after suffering from extended pneumonia, scours and/or chronic bloat. Chances that you will save it are very slim. Give the calf an isolated, warm pen with fresh feed and water. Don't spend too much money on additional medication.

Another pre-death symptom is temporary or permanent loss of eye muscle control, described as 'sky or star gazing'.

An otherwise healthy calf is suddenly found dead

Though many causes may be involved, lead poisoning is a likely suspect. Consumption of only 150–200 mg of lead represents a lethal dose. Sources of lead include discarded car batteries, certain herbicides, discarded paint tins or painted woodwork.

Calves that have ingested only small quantities of lead and are still alive look dejected, dull and have sunken eyes. They often show abdominal pain and grind their teeth. Treatment is available from veterinarians.

Visual changes in calves symptomatic of stress

The many types of calf scours

The characteristics of calf faeces can be a good indication of the type of digestive disorder being suffered. Here are a few examples:

Blood is present in the faeces of the newborn calf
The inner lining of the intestine of a newborn calf consists of immature cells that are replaced within a few days after birth by more permanent ones. During this period the fragile blood vessels can easily break. When the broken vessel is close to the end of the gut, bright red blood appears in the faeces. If bleeding is excessive, an injection of vitamin K (a blood coagulating agent) can be given. Otherwise the occasional appearance of blood in faeces should not be of great concern. When excessive bleeding is accompanied by high temperature and scours, coccidiosis or salmonellosis may be occurring.

It has white or yellow scours
This suggests that a number of the classic pre-scour signs such as loss of appetite, depressed appearance and facial hair standing on end were missed. Among the possible causes of the scours are inadequate colostrum, overfeeding, overcrowding, poor sanitation and stress in general. If the scouring was the result of inferior milk replacer, the volume of faeces will be unusually large with a gelatinous consistency.

It has watery scours
Mild cases of watery scours, usually lasting 6–12 hours, are often seen in purchased calves after about five days in the rearing shed. They are connected with the change in diet, stress or by slight overfeeding. Sometimes the faeces contain bloodstains originating from a broken vessel.

It has bloody scours and it is straining to pass manure
The presence of blood in the faeces may be of no significance or it may indicate serious infections from salmonella or coccidia. If a calf of 14 days or older has a normal or slightly elevated temperature, but its watery faeces contain large clots of fresh blood or dark tarry bloodstaining, it is likely to be suffering from coccidia. The infectious agent

is a common opportunist, present in more than 50% of healthy calves. The incidence of coccidiosis rises on farms where calves are subjected to early confinement and are exposed to massive infections at an early age. This is a stress-related disease and usually indicates a poor rearing environment.

It has loose, dark brown stools
This usually indicates bleeding from lesions and ulcers in the abomasum or serious infection in the digestive tract. When bleeding takes place in the abomasal area, medication seldom helps. Use gastric and intestinal protectants containing kaolin, pectin or bismuth. If the calf is eating solid feeds, reduce the acidity in the gut by feeding less grain and more roughage.

Its eyes are bulging

The eyes in some newborn and young calves protrude from the eye sockets, giving it an appearance similar to people with a thyroid disorder. Fortunately, bulging eyes in calves indicate a good supply of body fluids and a scour-free history. If the calf is healthy and attentive, bulging eyes should not discourage you from purchasing it.

It has droopy ears

This animal is likely to be running a high temperature because of pneumonia or a digestive disorder. Check the temperature and if it is high (above 39.7°C), the calf should be immediately treated with a broad-spectrum antibiotic. If one ear droops, check for external parasites such as lice and treat them with a few drops of hydrogen peroxide. If the base of the ear is swollen or tender, use a teaspoon of warm olive oil and gently work it into the skin. Any discharge from the ear usually indicates an infection that could be treated with antibiotics.

It has facial hair standing on end

When first observed in a previously healthy calf, this usually indicates an imminent digestive disorder. It is likely that the calf will be scouring within 24 hours. Skipping one milk feed (if twice daily feeding) and replacing it with electrolyte may help.

If the calf was purchased with facial hair standing on its end or if it is a permanent fixture, the calf has possibly had lengthy pneumonia and is still not feeling well.

It has sunken eyes and its skin has lost its flexibility

The problem is dehydration and it has not been recognised or treated for the last few days. Prolonged scouring leads to substantial loss of body fluids as well as electrolytes. The body of a young calf contains 75% water and a loss of 10% puts its life in danger, while a loss of 15% results in death. Sunken eyes are one symptom of dehydration, which, if advanced, will cause the upper eyelashes to be directed towards the inside of the eye socket, obscuring the calf's vision.

The level of dehydration can be checked by pinching a bit of skin near the ribs and twisting it 90 degrees. The slower the fold of skin springs back to its original position after release, the higher the level of dehydration and the quicker the need for treatment.

It has lost hair around its muzzle and/or rectum and along its hind legs

Offering hot milk to the calf or letting the manure stick to the skin for a long time are often stated as the main reasons for loss of hair. If these two causes can be excluded, poorly emulsified fat in milk replacer is a likely suspect. Fat globules attach themselves to the skin and prevent the air reaching the hairs. Similarly, hair is lost around the rectum when it is in contact with the undigested fat in the manure. Low digestibility of fat in milk replacers containing high levels of non-clotting plant protein may have the same consequences. If the flesh is raw, wash it with a clean cloth, wrung out with soda water. The whole area should be treated with a weak solution of iodine.

It bloats after drinking milk

Under certain situations the oesophageal groove does not close completely, thus leaking milk into the rumen. This can occur through rough handling, feeding milk that is too cold or too hot, overfeeding or force-feeding when the abomasum is not sufficiently empty. It can also occur when the calf is sick or when fed poor quality milk replacer.

Feeding milk through rubber teats, or at regular intervals, at body temperature and in small quantities may help re-establish the proper function of the oesophageal groove. Letting the calf suck your finger for a moment before offering the milk bucket will also help.

A weaned calf bloats on ad lib grain feeding

The sudden accumulation of gas in the rumen that cannot be expelled can even occur in calves that are well adjusted to high grain diets. Within one or two hours after feeding, the left flank rises very quickly; the calf nervously lies down and tries to defecate. If only one or two calves bloat, then it is unlikely to be due to the feed or feeding practices.

Some calves are just prone to bloat and will get over it without any treatment. Regrouping calves may allow previously 'bossed' calves better access to the grain, which can upset rumen gas expulsion if this happens too quickly.

A foul-smelling greenish liquid drips from its mouth and the calf loses its cud

This is sometimes called 'medicine disease' and is caused by prolonged use of antibiotics, which upset the balance of rumen microbes. The best option is repeated introduction of a cud from a healthy animal, preferably on the same diet, into the sick calf. A similar effect called 'microbe swapping' can also occur during hand feeding of calf starter in newborn calves.

Some rearers consider that the dripping is caused by calves twisting their heads to the side while drinking from rubber teats. This can cause the oesophageal groove to malfunction, thus allowing milk to enter the rumen and upset the establishment of normal populations of rumen microbes. Another possible cause is damage to a large portion of the rumen wall by prolonged scouring or the presence of small pieces of wire.

It develops a potbelly

This indicates a long-term nutrient imbalance in that there is too much fibre and too little energy in the diet. High fibre diets require high water intakes that, together with the slowly digested feed, increase rumen volume. Energy intake is further reduced through a limited gut capacity and this leads to poor growth. Sometimes potbellies develop in calves suffering from internal parasites, a damaged gut from chronic scouring or those with a long history of pneumonia.

The obvious treatment is to feed more energy and less roughage. By feeding ad lib concentrates and a low quality roughage, calves will only eat about 10–15% of their diet as fibre and the rest as concentrates.

The mouth cavity and skin under its eyelids are pale

Calves fed milk exclusively will show signs of anaemia. This is due to low iron levels in the diet. Once eating solid food this problem will disappear as concentrates contain sufficient iron for calf requirements.

If growing calves for white (milk only) veal, intake and performance can suffer through anaemia. This can be corrected, without endangering the marketability of calves, with intramuscular injections of iron.

It has a dry, hot muzzle

This calf would have a high body temperature and most likely be suffering from a respiratory disorder. Electrolytes and antibiotics would probably help.

It has a nasal discharge

A transparent, watery discharge indicates the calf is or was exposed to heavy environmental, housing or nutritional stress. The cause of the discharge is usually a viral infection, best comparable to a human cold. Remove the source of stress, and if the body temperature is elevated, offer three adult-size aspirins per day.

If the colour of the discharge changes to brown or greenish and is thicker, then the body is already fighting a secondary bacterial infection.

Its temperature dropped after treatment but rose again a few days later

This could be due to several possibilities:

- The correct treatment was applied but for too short a period.
- The treatment was applied once each day, whereas twice daily treatment would have provided better uniformity of antibiotic release.
- The level of drug applied was insufficient.
- A combination of the chosen antibiotic along with an anti-inflammatory drug was administered and the temperature drop was solely due to the anti-inflammatory drug, which may have masked the improper selection of the antibiotic.

It is important that the veterinarian should identify the disease organism responsible for the stress to ensure the most appropriate treatment can be given.

It has saliva running from its mouth

This may indicate many disorders but is mainly connected to a severe pneumonia. The front legs are spread, neck is stretched, the head points to the ground and breathing is laboured. Saliva will be running either in a clear steady stream or as a slow moving liquid. In many cases, even dramatic measures cannot save such a calf. Move it to a well-ventilated isolation pen, provide good bedding and fresh feed and water. Veterinary attention is essential.

It has an umbilical hernia

Hernia or rupture is a protrusion of one or two loops of intestine or other tissue from the abdominal cavity through the navel opening. If such an opening is no more than 2.5–4 cm, it usually closes sufficiently when the calf grows older. Larger openings require surgical correction.

Taping the opening for a period of four weeks may be necessary if it is two fingers wide at 2–3 months of age. Application of rubber rings (used for tail docking in lambs) to the skin pouch only, are effective in heifer calves. Use of more than one rubber ring prevents them from sliding down. The rings stop the blood supply to the navel and in two or three weeks, the navel cord will fall off and the connective tissue will close the opening.

It has warts

Warts are a specific skin overgrowth caused by a viral infection. In calves they sometimes appear on the head, the ears and around the mouth and eyes. They are contagious to other animals and some can even be transmitted to humans.

It has manure accumulation around its hooves

Dry clusters of manure can have very unpleasant consequences. They can cover an infection, be filled with fly maggots or lead to abnormal wear of the hoof. The feet should be checked at regular intervals using a blunt edge of a putty knife to remove the manure between and around the hooves. If the skin under the removed manure is red, mouldy or smells, wash it with diluted iodine.

Its mouth is cold

You are losing this calf. The body defences are breaking down and infection is taking over. The body temperature is well below normal, usually below 35°C and the chances of recovery are very slim.

In an attempt to raise its temperature, try thick, dry bedding, or plastic bags filled with warm water or heat lamps while lukewarm milk and/or water could be offered. Don't raise your hopes too high.

Understand how calves and heifers react to people

How much do we really know about the basic sight and hearing senses of calves and heifers? A recent article by two US calf rearing specialists (Sam Leadley and Pam Sojda, 2001) tells us much of what we may take for granted, but on the other hand, may not even be aware of. Firstly, cattle have wide-angle vision, if fact they can see 300 out of 360 degrees around them. They use this field of vision to define their 'personal space', which we call their 'flight zone'. Secondly, cattle are quite sensitive to high frequency noises compared to people (who can hear noises from 1000 to 3000 hertz), they can hear noises up to 8000 hertz. Leadley and Sojda have listed some general rules to help with cattle handling.

1. When a person moves into their flight zone, cattle will normally try to move away.
2. The size of their flight zone will decrease slowly if they are handled frequently and gently.
3. Previous experiences will affect how animals react to future handling, with memories persisting for many months. Obviously fear memories are significant in increasing flight zones.
4. Calves can readily tell the difference between two situations and make choices to avoid the more stressful one.
5. Cattle are sensitive to changes in colour and texture.
6. Moving objects and people seen through sides of a chute can frighten animals.
7. Novelty is a strong stressor, while repeated exposure will reduce the novelty effect.
8. Cattle are herd animals and don't like to be separated from their herd mates.

9. Groups of cattle that have body contact remain calmer.
10. Unexpected loud or novel noises can be highly stressful.
11. Cattle readily adapt to reasonable levels of continuous sound, such as background noises or music.
12. Cattle exposed to a variety of sounds, such as radios with talk and music, may have a reduced reaction to sudden noises.
13. Cattle readily adapt to handling, even if the events may be initially stressful, such as walking up a race, into a head bale or being transported.
14. Cattle can be trained to voluntarily accept restraint with relatively low levels of stress.
15. A small amount of inconsistency in care and handling can reduce calves' stress response to new sights and sounds.
16. Consistent poor handling can create chronic stress.

These basic rules can partly explain why empathetic calf rearers do a good job, whereas insensitive rearers do a poor job. Just spending time with young calves, particularly newborn ones, develops that essential bond, while quiet consistency in all management procedure, even to the point of clothes worn in the calf shed, ensures the calf nursery is as peaceful as any infants' bedroom.

Communicate with your calf rearer, too!!

Farm managers and other employers of farm staff should be aware that praise is one of the best motivators for employed labour. When your calf rearer does a good job, be certain to say frequently, out load and face to face, 'Thanks for doing a good job'!

Proficient calf rearing requires the use of the fives senses (sight, smell, hearing, touch and even taste) and this takes time to develop. Calf rearers have to be extra alert and ready to act quickly when a calf is ill. Timely diagnosis and treatment are measured in minutes rather than days. This kind of care calls for lots of flexibility and commitment on the part of the calf rearer. Also, good calf rearers have a bond with their calves that is tied to this commitment.

Giving that little extra over and over again, week after week, is costly for the rearer. It means being continually alert when working with the calves, so sick calves are quickly identified and treated. It may also mean returning to the calf shed at night to administer antibiotics or electrolyte fluids.

Owners and managers must allocate sufficient labour resources at the right time and place, for example, when assisting in parturition, such as dipping the navel with iodine and dosing newborn calves with colostrum. They should also provide opportunities to spread out stressful events rather than stack them one on top of another. For instance, vaccinations, dehorning, tail docking, ear tagging and weaning are all stressful to the calf, and to the rearer if it requires continual handling and restraining of calves.

Figure 9.2 Outdoor rearing can impose additional stresses on calves

When selecting or constructing rearing facilities, rearers should also be kept in mind. Ensure that staff can easily see all the calves during a single patrol down the calf shed. Provide enough hot water for cleaning buckets, teats and other feeding equipment, as well as cold water outlets for calves to drink from. Importantly, provide good quality and palatable feeds, such as concentrates, roughage and milk replacer. If calves quickly develop a taste for solid feeds, they require less labour input and are less likely to suffer ill health.

Skilled, motivated and empathetic staff are a major contributor to a successful calf rearing operation.

Contract calf rearing

Dairy farming is becoming a specialist profession requiring many skills. Rather than keep up to date with all these skills, increasing numbers of farmers now outsource particular enterprises on their farms. Contractors now offer their services in forage conservation, milking, heifer rearing and, recently, calf rearing. Skilled rearers collect newborn heifer calves from the farm, milk rear them and return the weaned calf at about 12 weeks of age, weighing 100 kg. The rearer is often provided with transition milk with which to commence milk feeding, but then the diet is changed to milk replacer. Early weaning, at, say, 5 weeks, would reduce total feed costs.

Recipient farms have to be selected carefully to minimise the introduction of diseases, thus reducing the rearers' concern about spreading diseases amongst the contracted calves. They should draw up formal contracts or at least agree on the costs of disease treatment and mortalities and target live weights. Rearers generally bulk-purchase milk replacer and calf pellets so they can develop good alliances with manufacturers. Current charges are about $15/calf/week or $180/calf for a fully weaned and healthy 12-week-old replacement heifer. As well as being 100% tax deductible for the dairy farmer, it would be compensated by the higher level of calf care than is often possible on seasonal-calving farms during the busy calving season. These rearers are also contracting with abattoirs and meat exporters to rear bull calves for southern Australia's developing dairy beef industry.

In some instances, calf rearers contract to rear the calves on the home farm, thus providing only the labour and expertise, therefore utilising all the farm facilities, including transition and vat milk.

There may also be a role for the specialist farm 'midwife', whose job it is to routinely check the springing cows and provide assistance with calving. They can assist with natural suckling of newborn calves or remove them at birth to sheltered pens, and artificially feed them their first colostrum. Such a position may only be needed for several months each year and that person could contract to do it on several farms in close proximity. The costs involved in employing such a farm midwife would be offset by the reduced time spent with the calving cows and their progeny (particularly late at night). There would also be the added benefit of reduced health problems and mortalities arising from a guaranteed higher level of passive immunity amongst the replacement heifers.

References and further reading

Drevjany, L. (1989), 'The Language of Calves', *Highlights in Agric. Res.*, Ontario, Vol.12, No.3, p.14. (This article includes only some of the signals mentioned above.)

Leadley, S. and Sojda, P. (1999), *Care and Feeding of Calf Raisers*, 'Calving Ease' newsletter, Dec. 1999.

Leadley, S. and Sojda, P. (2001), *Improving Heifer Handling (Parts 1 and 2)*, 'Calving Ease' newsletter, Dec. 2001/Jan. 2002.

ten 10

Disease prevention in calves

This chapter concentrates on the clinical signs of the most important calf diseases and on first aid and nursing in sickness and convalescence. It does not present a comprehensive catalogue of calf diseases, nor does it follow the pursuit of a diagnosis through post mortem examination, microbiology and clinical pathology.

There will invariably be calves that are either born dead or die pre-weaning. What constitutes an acceptable death rate? Roy (1990) considers that, under good management in developed countries, expected mortality rates are:

- Abortions (stillbirths <270-day gestation), 2–2.5%.
- Perinatal (stillbirths >270 days and during first 24 hours of life), 3.5–5%.
- Neonatal (between 24 hours and 28 days of life), 3%.
- Older (29–84 days), 1% or (84–182 days), 1%.

So excluding abortions, 7–9% of calves are expected to die between birth and 3 months of age. This seems to be the reported rate in the US, whereas in Australia it is generally lower at 2–4%.

The basic principles for good calf health are:

- Minimise exposure to disease pathogens.
- Delay exposure until calves can develop their own immunity.
- Maximise acquired immunity through colostrum and vaccinations.
- Keep calves sheltered, dry and free from stress.

The best way to maintain calf health is to ensure an adequate intake of colostral Ig within the first few hours of life. Good farm management should ensure this occurs (see Chapter 3). This is obviously difficult if relying on calf auctions or calf scales to supply animals for rearing. If calves have to be bought, it is preferable to buy them directly from the property of origin as this reduces their likelihood of picking up diseases in transit and the duration of stress and starvation. *Prevention* by adequate colostrum intake is far more effective than *cure* by drugs.

Rearing calves inside sheds at high stocking densities can provide an ideal environment for calf diseases to proliferate, although many still occur in calves reared at pasture. Prevention of future outbreaks through cleaning and disinfection is also more difficult when calves are reared in permanent fixtures. However, a warm, dry and well-managed calf shed usually offers better protection against diseases in young calves than a cold, windy and muddy calf paddock.

There are two major disease problems in calves in Australia, namely scours and pneumonia. These two account for more than 80% of all calf deaths, with scouring being the most common. Bloat, navel-ill, accidents and poisoning would make up the bulk of the remaining mortalities.

Calf scours or neonatal diarrhoea

The causes of scours in calves under 21 days of age are difficult to determine. There is usually not one single cause, but an interaction between calf management, diet, the environment, poor immunity and pathogenic viruses and bacteria can be factors.

Dietary scours

This mainly results from overfeeding (especially with cold milk) or incorrect milk replacer concentrations. Sudden changes in feed type, particularly changing from whole milk to milk replacer, or use of poor quality milk replacers, can also lead to dietary scours. Affected calves get severe diarrhoea but otherwise appear normal. However, they can more easily develop infectious scours. The best control measure for dietary scours is changing from milk to electrolytes for at least 24 hours. Some farmers and experts recommend taking calves off milk only as a last resort and then only after they are certain that an infective agent is the major cause of scours.

White scours

This generally occurs in the first few days and is usually caused by pathogenic strains of bacteria known as Escherichia coli (E. coli) that invade the gut wall. Foul smelling, grey to creamy-white severe diarrhoea is seen. Calves quickly become dehydrated and lethargic, will not eat, are 'tucked up' in the abdomen and may die suddenly. In chronic cases that linger on, infection of the lungs (pneumonia) or joints (arthritis) can occur. On post-mortem, a calf that died from E. coli scours will often show no visible signs of having an infection. Stress factors, such as cold or partial starvation (due to irregular feeding intervals as occurs when calves are sold in saleyards) can increase the occurrence and severity of white scours.

Viral and protozoal scours

These are generally caused by rotavirus or coronavirus (viral) or cryptosporidia (protozoal) and constitute most of the scours in calves less than 3 weeks old. Antibiotics do not kill viruses or protozoa and so are not effective in treating these scours. Furthermore, their overuse in treating scours will increase the risk of antibacterial residues in slaughtered bobby calves.

Salmonella scours

This occurs more commonly in older calves, causing bloody, putrid diarrhoea containing mucus. They develop fever, are weak and rapidly become dehydrated and emaciated. They have a high death rate. Less severely affected calves can have rough coats, potbellies and become stunted; they can also become carriers of salmonella and continually infect other animals. Extra personal hygiene is needed when treating salmonella, as the bacteria can infect humans.

Worm scours

These are caused by internal parasites eaten by grazing calves. These would not occur in housed systems unless purchased calves are older and have previously run at pasture. See the section on page 126 on internal parasite control.

Coccidiosis or blood scours

This is caused by protozoa infecting the calf from 3 weeks of age and onwards and can easily be confused with white scours. Affected calves show bloodstained scouring with a lot of mucus and may eventually develop anaemia. Coccidiosis is a stress-related disease and usually affects calves that are reared in wet, crowded, unhygienic conditions.

Treating scours

Scours accounts for 75% of all deaths under 3 weeks of age (Radostits and others 1994). The most important pathogens associated with infectious scours at different ages are:

- E. coli, 3–5 days.
- Rotavirus, 7–10 days.
- Coronavirus, 7–15 days.
- Cryptosporidia, 15–35 days.
- Salmonella, several weeks.
- Coccidia, older than 3 weeks.

Scouring calves can lose up to 20 times more fluid than healthy animals and they will become dehydrated because they are losing considerably more liquid than they can drink. This lost fluid also contains mineral salts and other nutrients. The degree of dehydration can be assessed using the skin fold (pinch) test. Pinch the skin and note how long it takes to return to normal. In healthy calves this is less than half a second. Another indicator is the degree of sunkeness of the eyes. Table 10.1 provides visual indicators of the degree of dehydration.

Table 10.1 Measures of dehydration in scouring calves

% Dehydration	Sunken eyes*	Skin fold test (seconds)	Clinical symptoms
4–6	–	1–2	Mild depression, decreased urine output
6–8	+	2–4	Dry mouth and nose, tight skin, still standing
8–10	++	6–10	Cold ears, unable to stand
10–12	+++	20–45	Near death

* The more pluses (+), the more sunken the eyes

Very dehydrated calves (10–15%) will require intravenous therapy. Calves with less than 8% dehydration and still drinking can be rehydrated orally by electrolyte solutions. Oral fluid therapy is the term used for treating scours with soluble sources of energy and electrolytes by mouth. These supply an energy supplement and replace lost vital minerals and fluids in scouring calves.

The amount of fluid required for daily maintenance requirements and to replace lost fluids can be calculated, based on live weight and the degree of dehydration. For a 40 kg calf with 6% dehydration:

- Replacement: 40 kg x 6% or 2.4 L fluid.
- Maintenance: 100 mL/kg/day, or 40 kg x 100 mL or 4.0 L fluid.
- Total: 2.4 + 4.0 or 6.4 L fluid.
- Feed this quantity in three feeds per day.
- Check the degree of rehydration using the skin fold test.

Up to 70% of calves will recover with adequate fluid therapy. Electrolyte treatments do not provide sufficient energy to maintain the animal. After 24 hours, reintroduce milk (if it has been withdrawn), but continue electrolytes for a further 48 hours. Separate milk feeding from electrolyte feeding by six hours. Rennet (junket) tablets added to the first two milk feeds will assist milk digestion.

The electrolyte solution should be offered to calves in the same manner as their milk (bucket or teat), but if they do not drink it this way, it can be administered using a drench gun or a stomach tube. It may be preferable to ask the veterinarian to give intravenous fluids to very sick and dehydrated calves because force-feeding often results in pneumonia since such weak calves cannot swallow properly.

Antibiotics may be required, especially if the calf remains dull after rehydration, and if blood appears in the faeces. Antibiotics must be used under veterinary supervision.

Prolonged use of antibiotics can lead to additional scouring because normal bacteria have been killed. This is known as medicine disease. Dosing the calf with plain non-pasteurised yoghurt helps re-establish abomasal bacteria (lactobacillus) used in milk digestion.

Veterinary advice should be sought to obtain an accurate diagnosis and the most appropriate treatment. Sick calves should obviously be isolated from healthy calves and tended to after feeding other calves to minimise the spread of infection. Water should be made freely available.

Diarrhoea powders containing kaolin, pectin or chalk or other methods of slowing down the passage of feed through the gut (such as charcoal tablets, corn flour or even sawdust) can reduce the severity of the scouring. Antibacterial compounds and antibiotics (for example, calf scour tablets, drenches or injections) should be used judiciously and restricted to cases where salmonella or other bacteria are suspected.

When prescribed, antibiotics are usually given orally for about three days. If the scouring is too advanced and the gut wall is badly damaged or the calf is running a temperature, a course of antibiotic injections may be required.

The infective organisms may be resistant to many disinfectants and survive in the environment for long periods. Formalin and hypochlorite are probably the most effective disinfectants, but only on well-cleaned floors and surfaces. Paddocks and yards are impossible to disinfect and require prolonged spelling. If possible, change the calf rearing area regularly as the risk of the disease is related to the build up of organisms. This is obviously easier if calves are reared outside in paddocks.

Controlling scours through management

Nutritional scours is caused by stresses reducing the production of digestive acids in the abomasum. Pathogens consumed by the calf are normally killed by the low pH from these digestive acids. If the acid production is reduced, then the abomasum does not protect the calf from these pathogens and they pass through into the intestines. The low acid production also reduces the effectiveness of the rennet in clotting the milk into a curd and so undigested milk then escapes into the intestines where it cannot be digested in the alkaline environment.

As the bacteria, normally resident in the intestines, now have a new supply of nutrients, they multiply and irritate the gut wall. This causes the body to secrete fluids into the intestines, thus leading to a loss of valuable minerals. Hence, a scouring calf becomes rapidly dehydrated and deficient in minerals. If this nutritional scours is not corrected promptly, the pathogenic bacteria that were not killed off by the stomach acids will also multiply in the undigested milk and the calf will develop infectious scours. By removing the initial stress, sufficient abomasal acids are produced, and normal milk digestion will eventually resume. Sudden changes in milk feeding routines are a common cause of scours. For example, calf rearers routinely report scours in calves about one week after changing from whole milk to milk replacer.

Environmental stress is another cause: sudden changes in weather or cold, damp and draughty or humid conditions inside calf sheds. Overcrowding is another cause, so sheds should never house more than they were designed to. Even changes in staff can lead to scours through different handling of calves, lack of TLC ('tender loving care') or changes in standards of hygiene. If reared outdoors, calves should always be offered protection against the extremes of sun, wind and rain. Despite this precaution, a sudden cold and wet spell can introduce sufficient stress to increase the incidence of scours in well-managed calves.

The duration of scours is largely under the control of the calf rearer. During their second week of life, calves are particularly susceptible. By careful observation, experienced rearers can anticipate the onset of scours the day before it happens, after which milk feeding can be reduced, with the calf recovering quickly.

The following signs of impending scours should be looked for:

- Dry muzzle.
- Thick mucus appearing from the nostrils.
- Very firm faeces.
- Refusal to drink milk.
- A tendency to lie down.
- A high body temperature (over 39.3°C).

Scours can occur under the best management but some precautions always help. If using a pad for calving down cows, calves should be quickly removed from any area used for holding these cows prior to calving to reduce the chances of manure contamination of newborn calves. This is also important in the prevention of navel infections and Johne's disease (refer to section later in this chapter).

A feeding routine should be quickly established with set feeding times, constant amounts of milk offered (and drunk) per calf and a consistent milk temperature. Any changes in feeding routine should not be too sudden. Individually pen newly purchased calves for the first two weeks, particularly if obtained from various sources, to quarantine them against the spread of disease to other calves. If buying calves from selected farmers, try to ensure that these farms ensure their calves get colostrum and have a low level of scours. Milk feeding equipment should be thoroughly washed and sanitised between feeds.

Clearly, early identification and treatment of sick calves is the key to their rapid return to health. Most scouring calves that are treated are back to normal after only two days on fluid replacer treatment and then they can be gradually reintroduced to milk over the next three days.

Some calf rearers include small amounts of disinfectant such as Dettol, or antibiotics in all milk fed. This may lead to low levels of infection in all animals, which only become apparent when calves develop more advanced symptoms. Furthermore, this practice can increase the growth of antibiotic resistant organisms, so making it harder to treat sick calves.

Many calf rearers have routinely used antibiotics to control potential pathogens, as well as to increase feed intake and utilisation. This is not necessary with ideal management and facilities, such as where colostrum intake is adequate, the rearing unit is clean and well ventilated and not densely stocked and the operator is experienced. Because this ideal scenario is not common, antibiotics have been used as insurance against disease, particularly when rearing calves bought from often unknown sources. This could mask any disease outbreak for several days and also give a false sense of security, which often leads to an even poorer job in calf raising. Concern about the development of antibiotic resistant strains of bacteria has led to the banning if this practice.

Figure 10.1 Separating milk-fed calves reduces the opportunity for disease spread

Preventing scours

To ensure healthy and disease-resistant calves, the importance of good colostrum feeding management cannot be overemphasised. Up to 40% of calves do not absorb sufficient antibodies into their bloodstream within the first 12–24 hours of life because of inadequate attention given to their colostrum feeding. Such calves are more likely to succumb to infectious scours. Chapter 3 discusses other aspects of colostrum feeding management, all of which can influence calves' susceptibility to scours.

Prevention of scours centres around good hygiene and minimising stress. Measures that can be taken include:

- Avoid buying calves from calf scales or sale yards, as these could introduce disease agents.
- Only buy calves directly from farms with good colostrum feeding management and good hygiene.
- Rest transported calves before their first feed of milk.
- Consider vaccinating cows for E. coli or salmonella prior to calving.
- Quarantine purchased calves for the first week or so, then disinfect the quarantine area after use, prior to introducing another batch of calves.
- Ensure calves are protected from extremes of climate, preferably in a shed.
- Carefully plan shed designs to avoid drafts and overcrowding.
- Minimise stresses associated with routine management practices, such as dehorning and castration.
- Maintain strict hygiene by cleaning and sterilising feeding utensils and facilities during milk rearing.
- Develop a routine milk feeding program, with as few people involved as possible.
- Develop an early weaning system to minimise the period of milk feeding.
- Quickly respond to early symptoms of scours, isolate sick calves and address the cause.
- Minimise the use of antibiotics and then only under veterinary supervision.
- Keep records of treatment of sick calves to assist in veterinary diagnoses and for withholding periods if the calf is subsequently culled.

Pneumonia and other respiratory diseases

Pneumonia is a problem with housed calves, particularly when stocking density is high and ventilation is poor. In the US, it accounts for 15% of the calf deaths from birth to 6 months of age. The shed temperature and relative humidity are the two most important factors influencing its occurrence. Respiratory diseases are more common in cool, damp sheds, although they can also be a problem in hot, dry shed conditions. Typical signs of pneumonia include lethargy, discharge from the nose and eyes, rapid breathing, and a rise in body temperature and pulse rate. Coughing is especially noticeable after exertion because of lung damage and affected calves are more susceptible to further outbreaks and secondary infections.

The control of pneumonia is mainly through improved housing. Poor ventilation leads to condensation, which results in humid conditions and an increase in the survival and spread of infection through water droplets in the air. Draughts of cold air at animal height in pens will aggravate the condition. Regular use of hoses in cleaning pens and laneways can introduce water vapour and blast infectious particles from the manure into the air. High dust and ammonia levels (the latter from urine in poorly drained pens) can cause irritations in the lungs making these calves more prone to pneumonia.

Early recognition and treatment of affected calves, with antibiotics, will minimise losses through deaths and poor calf growth. Sheds should be adequately ventilated but draught-free. The use of solid walls to at least 2 m high and then shutters or blinds to control air movement (particularly during cool weather) is ideal. In poorly ventilated sheds, well-positioned exhaust fans can improve air flow without causing draughts. Shed design is discussed in more detail in Chapter 11.

There are other, influenza-type respiratory diseases normally associated with high stocking densities in poorly ventilated sheds. These are often called 'crowding diseases' in Europe, for obvious reasons.

One such disease is infectious bovine rhinotracheitis (IBR). This is caused by a virus and leads to loss of appetite, fever and discharges from nose and eyes. The muzzle is often bright red (hence the name 'red nose' in Europe) and affected calves breathe with great difficulty. Like all respiratory diseases, secondary infections can confuse their initial cause and veterinary assistance is strongly advisable to ensure the correct treatment.

Pneumonia can also occur in grazing calves and lungworms can play a significant role in damaging the lungs. Adult worms lay eggs in the lung and these are coughed up, swallowed and then passed out onto the pasture. Larvae survive best in cool, wet conditions, so numbers build up on pasture in winter and early spring. Mature cattle have a strong immunity to lungworms whereas calves are very susceptible. Most drenches for round worms also control lungworms.

Pulpy kidney and other clostridial diseases

Pulpy kidney can occur when calves are first introduced to high concentrate diets. It is caused by one of the clostridia bacteria, which produces a toxin in the gut, eventually killing the calf (hence the name enterotoxaemia). As with all clostridial diseases, the bacteria are a normal part of the environment and are impossible to eradicate. The classical sign of pulpy kidney is that the fattest calves (the best drinkers) die suddenly and their carcasses rot very quickly. Routine vaccination programs of 'five-in one' vaccines can prevent the disease.

The other clostridial diseases controlled by five-in-one vaccines are blackleg, black disease, malignant oedema and tetanus. The dilemma with these diseases is that once you have vaccinated, it is hard to prove that it has been worth it – you do not know if you would have lost calves if you had not vaccinated. However, the vaccine is cheap and the cost of a vaccination program is negligible compared to the potential losses incurred through clostridia. It is important to follow the manufacturer's instructions, with regard to age of initial and booster vaccinations. A combined 'seven-in-one' vaccine provides protection against both clostridia and leptospirosis diseases.

Calves that have drunk sufficient colostrum soon after birth can be partially protected against clostridia up till 6 weeks of age, after which a vaccination at 6–12 weeks of age with a follow up one at least six weeks later, gives good immunity. A booster vaccination 12 months later should reduce the incidence of clostridial diseases in adult cattle and this should be repeated every 3–4 years. Deaths from clostridia have occasionally been observed following complete vaccination programs, which means that immunity is not always complete.

Internal parasites and their control

Round worms and liver fluke are the two most important internal parasites that require attention. Round worms cause gastroenteritis in young weaned calves. The intestinal worms damage the gut lining, decrease appetite and interfere with the efficient absorption of nutrients. The signs are scours, weight loss, bottle jaw, dehydration and sometimes death. Mild signs include ill thrift and a dirty tail.

Calves pick up infective larvae while grazing and these mature in the gut within two–three weeks, mate and start to lay eggs. The most troublesome round worm in the southern, winter-rainfall regions of Australia is the small brown stomach worm (*Ostertagia*). Barber's pole worm (*Haemonchus*), nodule worm (*Oesophagostomun*) and hair worm (*Cooperia*) are the major round worms affecting calves in the northern, summer-rainfall regions of Australia. There are seasonal peaks of worm burdens that should be considered when planning drenching programs. Mature cattle are relatively resistant to round worms while young stock are the most susceptible.

Worm control depends, first, on strategic drenching to suppress worm burdens and to prevent contamination of pasture by worm eggs and, second, on integrating drenching with grazing management. Drenching only kills the worms in the calf and does not prevent reinfection. Housed calves have not picked up worms and, hence, do not need drenching.

Drenching programs vary with the area and local recommendations should be followed. Worm test kits are now commercially available and these can assist with parasite control programs, particularly in determining which drench will be the most cost-effective.

Liver fluke depend on a fresh water snail for its lifecycle. Acute fluke disease results from massive damage to the liver, caused by the immature flukes, and can kill calves. Chronic fluke disease is due to the adult fluke blocking the bile ducts of the liver and can lead to weight loss, anaemia, bottle jaw and scours. Adult cattle build up resistance to flukes. If control of snails or control by grazing management is not possible, for instance, in irrigation areas, drenches can be used to remove adult and immature flukes before snails become active in the warmer weather. As with all drenches, it is important to read the labels for dose rates, warnings and withholding periods.

There are other types of worms or internal parasites likely to infest calves in different areas of Australia and local advice on drenching and other control measures should be sought.

Johne's disease

Johne's disease is an incurable bacterial infection of the intestines that is known to occur in more than 13% of Victorian dairy herds, but much less in other states. By the time clinical symptoms develop, the wall of the intestine has become thickened and this interferes with the absorption of nutrients from the digested feed. Cows with Johne's disease show progressive chronic diarrhoea and weight loss, ending in death. However, they generally remain bright and alert and maintain a good appetite up to the time of death.

Most infected cows will not show any signs of the disease and stress is important in determining whether the symptoms appear in infected cows. Stresses may include calving, cold weather and feed shortages or moving cows to a different herd or farm. Once Johne's disease is detected in a herd, it is usually well established and there are likely to be other infected carrier cows.

Apparent freedom from clinical cases, even for years, provides no assurance that the herd is free of the disease. Infection occurs during calfhood, but the symptoms are not usually seen until infected cattle are 4–5 years old. Cattle become resistant to infection by about 12 months of age.

The disease is spread by a susceptible calf consuming feed, water or milk contaminated by manure from an infected cow. Occasionally, an unborn calf can acquire an infection from a diseased cow.

Control of Johne's disease depends on separating calves from their dam within 12 hours of birth and then rearing them until 12 months with no contact with faeces from adult cattle. Attention should be given to paddock drainage and not applying reuse water when irrigating calf or heifer paddocks. Drinking water should be supplied from clean sources via troughs, not dams or irrigation drainage channels.

The infective bacteria can survive for up to 12 months in cool, moist conditions. They are destroyed by sunlight and dry conditions.

Johne's disease is a notifiable disease in all states. All cattle showing clinical signs should be reported to local animal health or veterinary advisers. Any herd with a history of the disease now enters a control program supervised by government veterinarians. Control programs vary from state to state, with some aiming for eradication. Infected or reactor cattle should only be sold for slaughter. The Johne's disease status of the herd affects the ability of cattle to move between zones, states and countries.

Victorian dairy farmers are encouraged to participate in a Johne's Disease Calf Accreditation Program (NRE 2000), details of which are available from government animal health and veterinary advisers. Complying with this audited quality assurance program accredits farms that rear their calves to minimise the risk of contracting Johne's disease. It does not guarantee Johne's disease-free cattle, but when undertaken, the program provides calves with much lower disease risk than for calves from non-participating herds. There is also a coordinated national market assurance program for Johne's disease, CattleMAP (AAHC 2000).

Preventing Johne's disease depends on two important factors: firstly, stopping the spread within a herd by good calf rearing practices; and, secondly, stopping the spread of infection by sourcing replacement heifers from low risk herds. Calf rearing systems should prevent calves ingesting feed or water contaminated with manure, while a closed herd is the most effective method of avoiding the introduction of the disease to a herd. Calf rearing practices should endeavour to:

- Calve cows in a clean paddock, as calving pads or heavily stocked calving paddocks presents a very high risk of spreading the disease.
- Separate the calf from the cow within 12 hours of birth.
- Ensure no cow manure or dairy effluent comes in contact with calves or the calf rearing area.
- Feed calf milk replacer or milk from low risk cattle.
- Only supply tank, town, or bore water to calves up to 12 months of age. Avoid stock dams, troughs or discharge/reuse irrigation channels.
- Prevent any adult stock entering the designated calf rearing area. This includes bulls, dry cows, milkers, sick cows, steers, goats or camelids (alpacas, llamas).
- Maintain strict hygiene when entering the calf rearing area. Do not introduce dung on boots, clothing or farm machinery such as tractors and bikes.
- Fence off the calf rearing area from laneways and milk tanker tracks.
- Maintain accurate records of calves reared or purchased.

After the calves are weaned and up to 12 months of age, the risk of them becoming infected will be minimised if:

- Weaned calves only graze paddocks that have had no adult cattle on them for at least 12 months.
- This grazing area is free of any drainage, effluent or sprayed recycled effluent and discharge/reuse irrigation channels are fenced off.

- Stockyards that are used by adult stock are not used by the calves.
- Calves sent on agistment are only mixed with stock that have had the same high Johne's disease rearing standards and only graze on areas free from potential contamination.

The degree of compliance by dairy farmers in Victoria to many of the above recommendations was assessed in a recent survey (Wraight and others 2000). They found that, even though 13% of the 540 farmers surveyed stated that Johne's disease had been diagnosed on their farm, there was a low level of compliance. For example, only 14% of the farmers minimised potential contamination of newborn calves from dairy shed waste or run-off from possible infected paddocks during birth and only 14% removed them from their dam within 12 hours. Although 38% minimised exposure of calves to manure, only 1% used separate boots and aprons when working in calf areas. Only 42% of farms excluded adult cattle from the heifer paddocks. Farms with a history of Johne's disease did make more of an effort to reduce exposure, but 91% of the farmers failed to comply with more than two of six recommended control measures.

Other diseases in calves

Bloat or tympany

Bloat is an over distension of the abomasum or rumen due to the gas produced by normal fermentation of feed being unable to escape. It can occur in the abomasum of calves fed milk and the rumen of calves fed milk, concentrates or pasture. It is more likely to occur where calves do not suckle the milk and where too much milk is fed too quickly.

The feeding of chopped straw seems to overcome these problems except where the bloat is caused by an obstruction in the throat or oesophagus. Animals can often show signs of bloat following feeding but the gas will escape and the rumen or abomasum will eventually return to normal size before the next feeding. If this does not occur then the use of a stomach tube or a trochar to relieve pressure in the rumen is recommended and veterinary advice should be sought. Prompt action is essential because affected calves can die within an hour after feeding.

Abomasal-induced milk bloat occurs when partially digested milk from a previous feed is enveloped in a clot in the abomasum together with newly drunk milk. Any gases being produced from this partially digested milk cannot escape, causing distension of the abomasum. Rumen bloat can occur in the calf fed milk or milk replacer through failure of the oesophageal groove to close properly or due to back flow into the rumen from the abomasum. This is particularly associated with diets containing certain non-milk proteins that are rapidly fermented in the rumen.

Abomasal-induced milk bloat appears to be more prevalent with certain types of milk replacers, particularly those incorporating tallow as an energy source. One experienced calf rearer includes bloat reducing chemicals such as terric with the milk replacer at every feeding.

Bloat in grazing, weaned calves is the result of a stable foam developing in the rumen, which traps the bubbles of gas produced by the rumen microbes. The foaming agent is present in the leaves of certain legumes, such as clover and lucerne. Treatment is urgent and affected animals can be drenched with 150–200 mL of vegetable or mineral oil, or even butter, lard or cream. If bloat is so severe that the animal cannot swallow, the oil can be inserted directly into the rumen, on the left side, using a wide bore needle. Leave the needle in place to allow some gas to escape. A sharp knife or trochar should be used as a last resort. A small vertical stab wound (2–3 cm long) can be effective and will heal faster than a large hole. Oils, detergents and monensin in anti-bloat capsules can also be used to control pasture-induced bloat.

Feed toxicities

These can occur through human errors in preparing feeds, supplying inappropriate feeds for calves or providing access to poisonous items in rearing sheds or at pasture. In one instance, Heliotrope poisoning was diagnosed in early weaned calves through contaminated grain being used in the concentrate pellet. In another instance, a producer suffered calf losses through incorrect levels of antibacterial drugs being incorporated into commercial feed preparations.

Calves are very susceptible to lead poisoning and this has occurred through animals licking or chewing painted woodwork and metalwork, discarded paint tins batteries and painted tarpaulins.

Gossypol, which naturally occurs in cottonseed, can poison calves. The heat process usually destroys it during extraction of the cottonseed oil. Mature cattle are not affected by gossypol (hence, they can be fed whole cottonseeds) as it is broken down in the rumen. Young calves cannot tolerate it. Cottonseed meal is generally very low in gossypol but for safety sake, it should not constitute more than 20% of calf grower rations.

The incorporation of soya flour in milk replacers can create problems if it has not been heat treated to remove the trypsin inhibitor. Trypsin is involved in digestion of milk in the abomasum. Soya flour (like other non-milk protein sources) cannot be utilised by calves less than 3–4 weeks of age.

Vitamin A deficiency has been diagnosed in calves that were rapidly growing and were not early weaned off milk replacer. These calves had low body reserves of vitamin A, were reluctant to eat concentrate pellets and were only offered limited levels of milk replacer.

Pasture toxins can be a problem in certain regions. Ryegrass staggers can occur during autumn in southern Australia, although this rarely causes deaths. Bracken fern poisoning can kill calves through damaging the blood forming tissue in bone marrow. Calves pass blood from their rectum, nose and mouth and respond very poorly to treatment. Deaths have been reported for up to six months after calves have been removed from the area.

Grain poisoning or acidosis

This is the result of rapid fermentation of cereal grains and other high starch feeds in the rumen, leading to excess levels of lactic acid being produced. Affected calves become dull and refuse food, their movements are unsteady, they often scour, and bloat may occur. Mild acidosis can be treated by drenching calves with a sodium bicarbonate solution and feeding a roughage-based diet, but severe cases require veterinary attention, as death can be sudden. The routine use of sodium bicarbonate and other rumen buffers when feeding high levels of grain should maintain normal levels of acidity in the rumen. The feeding of chopped straw will stimulate saliva production, which buffers the rumen against rapid changes in acidity.

Navel-ill and joint-ill

This is caused by bacteria infecting the umbilical cord soon after birth, particularly where the calving area is heavily contaminated. Unless treated promptly in young calves, it can lead to severe inflammation or arthritis of the joints. Animals with joint-ill are reluctant to walk and stand for only brief periods. As the infection is carried in the blood stream to all parts of the body, reduced appetite, diarrhoea and pneumonia may also occur. Navels in all newborn calves should be swabbed or sprayed with diluted iodine (7%) as a precautionary measure, and calving facilities should be kept clean. This and other navel abnormalities (such as umbilical hernias) should be apparent when selecting calves for purchase; these animals should be rejected.

Pink eye

This is a bacterial infection of the eye, which occurs mostly in the warmer months, possibly as a result of increased fly activity and dust and irritation from young grass. Calves and young stock are more commonly affected than older cattle. The first sign is a discharge from the eye, then it becomes reddened, a shallow ulcer develops and finally the eyeball looks white. Most affected animals recover, leaving small scars that do not appear to interfere with sight. In some cases, the eyeball can rupture and blindness results.

Treatment with various ointments, powders or sprays has little effect unless the level of antibiotic is maintained at a high level by frequent applications. Severe cases should be protected by gluing a patch over the affected eye or suturing the eye shut. Control is difficult because little can be done to avoid exposure to ultraviolet light and dust. Fly control helps but isolation of clinical cases is not effective because normal cattle carry the causative bacteria.

External parasites

Flies breed in manure and moist feed waste, so these should be removed regularly. Biting flies can cause worry amongst calves so standard fly control measures (such as fly bait or routine spraying) may be necessary. Lice control using ectoparasite dips or sprays may also be occasionally required due to severe lice infestations, particularly in poor calves. Ringworms can occur and should be treated with anti-fungal preparations. Some external parasite treatments should not be used simultaneously with worm drenches while others should not be used on young calves. It is important to carefully read the instructions before use.

Cattle tick and buffalo flies are a problem in northern Australia. They suck blood and cause skin irritation and also can carry potentially fatal diseases such as tick fever. Therefore, it is important to spray or dip calves routinely during the tick season and also to implement recommended grazing management procedures, such as pasture spelling and rotation.

Leptospirosis

This is a bacterial disease that can occur in all farm animals. The lepto bacteria gain entry through the skin or membranes lining the nose, eyes or mouth, or by ingestion. They multiply in the liver, enter the bloodstream, and settle in the kidneys. They are then passed out in the urine. The two most common types of leptospirosis bacteria that affect cattle are *Leptospira pomona*, which can cause abortion, mastitis and milk production losses in mature cows, while it can cause redwater (blood in the urine), jaundice, anaemia and death in calves.

The other bacteria, *Leptospira hardjo*, seldom causes disease in cattle but it does affect humans who catch it from the urine of cattle. Humans show flu-like symptoms including fever, chills, headache, muscular aches and vomiting. About one in 10 dairy farmers are likely to acquire 'lepto' as an occupational disease in a working life of 30 years at the rear-end of cows.

The best way to reduce the risk of milkers acquiring lepto infection is to vaccinate all heifer calves and cows. This will stop cows shedding the bacteria in their urine in the dairy. Calves should be vaccinated twice, about a month apart, once they reach six months of age. Combined lepto and clostridial vaccines, seven-in-one, are the usual form of vaccination. A booster vaccination should be given 12 months later and followed up by an annual booster at the time of drying off. In-calf heifers are a high-risk group and should always be vaccinated.

Depending on the area, other vaccinations or preventative measures are advisable with young calves, such as for clostridia or for tick fever.

How to recognise sick calves

Before rearers can recognise sick calves, they must know how healthy calves behave. This allows them to be on the alert for subtle changes in calf behaviour before clinical signs of disease become obvious. They should never be complacent about changes in calf well-being and behaviour.

Figure 10.2 Calves reared in large groups require good surveillance for disease control

Calves charging your knees and running around the pen are healthy. Such calves rest in a curled-up position with feet tucked under and heads back along the body. They appear relaxed with regular breathing. Some healthy calves may also rest flat on their sides.

Signs to look for

Each day look quickly over each pen of calves, then be more specific and check suspect calves' noses for dampness and ears for temperature. Sick calves often have dry noses, higher than normal body temperatures and a depressed attitude. Listen to their breathing, noting any 'rattles' or laboured breathing. Lift their tail and note the state of any faecal residues. Look at their feet and legs. For the first week to 10 days of age, check the navel area for signs of inflammation and swelling. This inspection should be undertaken as part of your daily routine.

Calves resting in the corner of pens, with their head turned away from pen-mates should not be ignored. Get the calf up. If it stretches, it is okay. If it does not, it may require further attention. Sick calves show general disinterest, become listless and apathetic, lack vigour and often do not move when approached. They may stand with their ears lowered and head down.

Calves must be kept in a stress-free environment. It is difficult to identify changes in the behaviour if calves are kept in conditions where they look miserable and hunched up because of cold stress.

Keep records to help identify problems

Records should be kept of changes in the intake of milk and concentrates and of fluctuations in growth rates. Body temperatures should be recorded in suspect calves to assist with disease diagnosis.

The reasons for outbreaks of scouring must be tracked down. It may simply be due to a change in feeding or management routine, in which case little further treatment is necessary. However, if scours persist, veterinary diagnosis should be sought. Only use antibiotics under instructions from your veterinarian.

Calf management and disease

Disease problems with calves seem to be worse during winter and are more frequent in calves with low birth weights. Calves become more resistant to diseases as they get older.

Scouring is more of a problem in milk-fed calves and in group penned rather than individually penned calves, and also seems more common in Jerseys than in Friesian calves. Scours are more prevalent in calves fed milk replacer than whole milk, but this could be related to aspects of mixing the replacer or its more variable quality, rather than any more healthy attributes of whole milk. Feeding the youngest, more susceptible calves first each time will minimise any disease transfer from older animals.

Pneumonia, on the other hand, is more prevalent in older calves (6–8 weeks of age) and is not affected by group or individual penning. Early weaned calves seem more susceptible to pneumonia than those fed milk for a longer period. Purebred Friesians also seem more affected than Hereford x Friesian calves. Scouring and pneumonia are more of a problem in calves purchased from calf auctions than those home bred or purchased directly off farm. Sex and weight for age have little influence on incidence of these diseases.

Diseases are more likely to occur in calves subjected to stresses than if adequate attention is given to their physical and nutritional needs. Examples of stress include lengthy transport from calf auctions in overcrowded, unprotected trailers or weaning off milk before the rumen is fully adapted to dry feeds. If there is a smell of ammonia in the rearing unit, better ventilation and/or floor drainage is required to reduce the likelihood of pneumonia outbreaks.

The best indicator of health is body temperature. Normal temperatures are 39°C (103°F) in the morning and 39.2°C (103.5°F) in the evening. When the body temperature rises higher, close examination and, often, treatment are necessary. Body temperatures are easily taken with a thermometer in the calf's rectum for one minute; 20 cm of string attaching the thermometer to a paper clip (which can be clamped onto the calf's hair) should prevent breakage. Electronic thermometers give a meaningful reading within five seconds.

It is up to the calf rearer to decide whether to do nothing, to treat the animal themselves or to contact a veterinarian. The other essential step to take if the disease is infectious is to stop the spread the disease by moving the sick animal to an isolation or 'hospital' pen. For all diseases, diagnosis should be obtained and post mortem reports should be sought for any unexplained calf deaths. Veterinarians can send samples to local veterinary laboratories for pathological examinations.

Like any professional, most veterinarians have specialised interests in their disciplines. When attending dairy farms in the US, veterinarians spend less than 5% of their time working with pre-weaned calves. It is not uncommon in intensive dairy regions for farmers with calf rearing problems to seek assistance from a particular clinic and a particular veterinarian because they have been happy with the way they have previously worked together. When rearers employ a veterinarian, they should not be afraid to follow him or her around the shed asking questions about the techniques used for diagnosis. When clearly explained, most of the techniques are obvious in solving the disease problem encountered, which means that in the future, rearers could undertake most of the diagnoses themselves.

When commencing a new calf rearing operation, or reassessing an existing one, it is always a good idea to find a veterinarian that you can work with. It is important to develop a disease prevention and management program, before any problems arise. Good calf rearers should rarely need a veterinarian. They should be able to identify the early signs of ill health then act on them before the calf requires much treatment. In my experience, the best calf rearers are women, because of their more empathetic and caring natures, and the best women calf rearers are hospital nurses, because of their training to anticipate health problems before they occur.

Being closely managed, most calves respond whenever people enter rearing sheds. The first sign of disease is general disinterest, in that calves appear listless, apathetic and lack vigour and will not move when approached. Calves standing with head down and ears lowered are likely to be showing early symptoms of disease. If calves do not stretch when standing after a lengthy rest, they should be carefully observed for other signs of ill health. Loss of appetite, dry and dull coat, sunken eyes, runny eyes and/or nose, fever and difficulty in breathing are obvious signs of ill heath.

Occurrences of disease and deaths are generally lower on farms where owners rather than employees rear the calves. Furthermore, deaths are lower on farms where the farm wife rather than the farm husband cares for the calves; death rates on farms where the children rear the calves are intermediate between those of employees and the wife or husband. Correct calf rearing, one of the most arduous tasks on dairy or beef farms, requires a genuine concern for the welfare of each animal (in other words, a sense of caring) and a quick recognition of the early symptoms of the diseases described above. One common attribute of all successful calf rearers is TLC. Young calves give out many signals indicating their health and general wellbeing and recognition of these signals becomes second nature to calf rearers with TLC (see Chapter 9 for more details).

What should you do with sick calves?

Unhealthy cows cost money. They have higher drug, veterinary and labour costs, and reduced performance, that is lifetime milk yield and number of calves born. Not only do unhealthy cows cost you money through fewer lactations in the herd, their higher culling rates increase the need to rear more replacement heifers.

What, then, should you do with sick calves? Our inherent nature is to provide them with all the TLC, veterinary assistance and drugs required until they are up and about running with their pen-mates. But what then? Should you keep them and grow them out or sell them at the first opportunity?

How much do sick calves cost? It is relatively easy to record the cash costs of treatment, such as veterinary visits and drugs. It is more difficult to cost out the extra time and care required during treatment and recuperation. For example, US researchers found that each sick calf required, on average, 53 minutes of extra care before recovery occurs. However, it is the long-term effects on heifer health and subsequent performance that are near impossible to quantify. These are much higher than the costs and labour during treatment.

Overseas studies have consistently found that sick calves have poorer performance as adult cows. For example, in Canada, heifer calves that were treated for scours were two to three times more likely to be sold prior to mating and three times more likely to calve down as 30-month-old, rather than 24-month-old, heifers. Furthermore, those that were treated for pneumonia during their first three months of rearing were two to three times more likely to die within this 90-day period. Such problems are reflected in wastage rates, which we can describe as the proportion of live heifer calves born that are culled or die before their second calving. Overseas targets are for 20% wastage, whereas Victorian surveys have recorded wastage as high as 35%.

So when should you decide whether to sell a calf or not? How sick should she be before you have to decide that she is never likely to be a really profitable member of your milking herd? There is no easy answer to this quandary. I suppose all we can conclude is that the more attention a sick calf requires during treatment, the less likely she will make you money as an adult cow.

Some astute farmers are adamant that every sick calf should be disposed of, either by humane slaughter or sale as a cull. The problem with selling such animals is that, unless the animal goes straight to the abattoirs, potential poor growth is just passed onto the new purchaser. However, one would expect that an astute calf rearer is unlikely to purchase a recovered calf.

How many farmers actually document which calves get sick, the degree of treatment required for their recovery, then their age and reason for culling?

If this record keeping became routine, farmers would then know how many lactations such animals are likely to remain in their milking herd. This could eventually provide a valuable benchmark for them to make the decision to cull recovered calves or let them join their healthier heifer calf mates. Only then can farmers make truly objective decisions as to the fate of their previously sick heifer calves.

Maintaining a healthy calf shed

It is important that calf sheds be maintained as disease-free as possible. The main avenues for introducing new diseases are via calves purchased off-farm, contamination from adult stock on-farm (such as Johne's disease) or from service providers visiting the calf shed. Biosecurity is the term used to describe the restricting of access by other livestock and personnel. One example of biosecurity, at a national level, was the restrictions of employment of staff on US dairy farms who had previously worked on dairy farms in Europe, following the 2001 outbreak of foot and mouth disease. Furthermore, Australian veterinarians who assisted with controlling this disease outbreak in England were required to not visit farms for at least seven days after their return.

One US county has recently formed a group of concerned dairy professionals to categorise service providers into high and medium risk, based on their contact with livestock and farm equipment. This approach could be used for calf rearing operations.

In that case, high risk people would include veterinarians and other calf consultants, operators of mobile calf scales, drivers of feed trucks, dead stock removal personnel, visiting farmers (both local and from other areas) and sales representatives and service personnel (with access to other calf operations). Medium-risk people would include consultants, sales representatives and service personnel (without access to other calf operations) and non-farm visitors.

The committee suggested the following protocols for high-risk service providers:

- Wash and disinfect boots before and after visiting the calf shed.
- Park vehicles away from the shed.
- Use clean and disposable coveralls at all times.
- Wash and disinfect all equipment.

- Collect and deliver livestock in clean trailers and trucks.
- Pick up dead stock away from the calf shed.

Medium-risk service providers could be asked to follow the first two of these protocols.

Other actions that could be taken are:

- Place signs such as: 'Biosecure area – do not enter without permission' at the calf shed and 'No visitors, sales or services people allowed without an appointment' at the front gate. This would necessitate listing a contact person and telephone number.
- Feed delivery personnel should wear plastic boots and be encouraged to use non-returnable bags.
- Place the bulk feed bins away from the calves, if possible.
- Ensure that all calves purchased off farm are shedded separately to the home calves, at least until they are 2 weeks of age.

While not all of these measures are practical on all farms, they should at least be considered, and some should be implemented to reduce the risks of introducing new diseases into your calf shed.

References and further reading

Australian Animal Health Council (2000), *The New Market Assurance Program for Johne's Disease*, Canberra.

Hides, S. (1992), *Dairy Farming in the Macalister Irrigation District*, Second Edition, Macalister Res. Farm Coop., Maffra, Victoria.

Jeffers, M. (1999), *Johne's Disease in Cattle – Prevention*, Agricultural Note AG0845, NRE, Melbourne.

Natural Resources and Environment (2000), *Johne's Disease Calf Accreditation Program Manual*, Melbourne.

Radostits, O., Leslie, K. and Fetrow, J. (1994), 'Health Management of Dairy Calves and Replacement Heifers', *Herd Health – Food Animal Production and Medicine*, Second Edition, Chapter 8, p.183–227, W.B. Saunders Co., Philadelphia, US.

Roy, J. (1990), *The Calf, Vol.1. Management of Health*, Fifth Edition, Butterworths, Sydney.

Schrag, L. (1982), *Healthy Calves, Healthy Cattle*, Verglag l. Schober, Auenstein, West Germany.

Wraight, M., McNeil, J., Beggs, D. (plus nine others) (2000), 'Compliance of Victorian Dairy Farmers with Current Calf Rearing Recommendations for Control of Johne's Disease', *Proc. 6th Int. Coll. Paratuberculosis*, p.157–168.

11 Housing of calves

Exposure to cold, wet and windy weather conditions can cause deaths, particularly with weaker animals. Calves and yearling heifers given access to shelter will grow more rapidly during winter than calves with no shelter, particularly if the sheltered area also provides a dry surface on which to lie.

Some form of shelter is needed, at least for young calves. Shade trees may be good on hot days but hedges or tree shelterbelts are not really sufficient during winter. Calf sheds need not be expensive structures but they should be built in ways that allow for easy cleaning and maintenance of hygienic conditions. Sheds should be at least twice as deep as they are wide or high to be draught-free, even with the front open. They should be well ventilated, without draughts and, ideally, open to the north or northeast to get the maximum winter sunlight without the prevailing wind and rain. If ventilation is sub-optimal in a galvanised iron shed – it is easy to create extra draughts, above calf height, by making shutters from some of the solid wall.

Sunshine provides a number of benefits to calves as it warms calves, dries out and disinfects their bedding. It also stimulates vitamin D production (which is important in utilisation of milk). The shelter is considered adequate if it is comfortable for a person to sit dressed only in light clothing. Calf housing should be directed more towards protection against severe extremes rather than from normal seasonal variations in which the calves' hair coat and natural instinct provide protection.

Types of shelter

There is a wide diversity of shelters used for rearing heifer replacements on dairy farms. Some of the more common ones are:

Small, galvanised iron sheds

These are cheap and simple and can either be permanent or on skids to move from one calf rearing paddock to another.

Converted hay sheds

These usually contain simple individual or group pens with deep litter floors on a clay base. They may or may not be enclosed on several sides.

Greenhouse barns

These new innovations in calf rearing are sheds made from steel tubing frame in a hoop covered with clear or translucent plastic. The sides can be rolled up to improve ventilation, while shade cloth or black plastic can assist with excluding the sun in summer.

Rows of calf hutches

These are sometimes called calf motels and consist of hutches placed side by side, each one enclosed on three sides. The design aims to provide ample ventilation without exposing calves to direct drafts. Each hutch has a concrete floor, maybe with some deep litter or a wooden grating.

Small individual calf hutches

Expensive fibreglass hutches are imported from North America, but they can be purchased locally quite cheaply ($70–$100 each). These consist of a galvanised iron 'A-frame' with a weldmesh enclosure, sufficient for the calf to move around in front of the hut. Large fruit packing cases are also suitable, provided they give sufficient protection from cold draughty winds. Calf hutches can be moved to new areas with each batch of calves to reduce disease build up.

Specifically constructed enclosed calf sheds

These are becoming popular as dairy farmers pay more attention to management of their replacement heifers. They vary greatly in design but would have solid floors and walls with shutters or blinds for ventilation. Further design features are described below.

In Australia, there is little need for the fully enclosed, insulated and temperature controlled calf houses frequently seen in Europe. Such sheds have been found to increase pneumonia in calves during winter, when compared to more open sheds and calf hutches, despite the freezing conditions.

Management considerations

A good calf house is one that meets the demands of both animals and operator at reasonable cost. Calves obviously need protection from rain, draughts (in winter) and direct sunlight (in summer), as well as a clean, dry floor on which to lie. The operator needs a building that is comfortable and convenient to work in, particularly for routine daily tasks such as feeding and cleaning. The operator should also be able to clearly observe all calves at all times (during day and night) and be able to catch, single-handed, restrain and administer treatment to any animal requiring attention.

The building and fittings should be designed to allow for easy cleaning and disinfecting between batches of calves. The shed should provide adequate space for storage of feed, hospital pens for sick calves, a desk for record keeping and a washing-up area (with hot water) for milk feeding utensils. Protection against vermin and birds, and possibly even flies, is important. The siting of the feed store and the location of doors and passages are important, as these will greatly influence the total distance covered each day. Therefore, the shed should be designed to enable tasks to be carried out quickly and without unnecessary repetitive movements. These aspects, together with detailed designs of rearing shed and pen layouts, are fully covered in the UK book *Calf Housing Handbook*, which is now available in Australia.

Pre-weaned calves are more susceptible to disease and cold stress than older animals. They require more intensive management such as milk feeding and, ideally, individual stalling. It is logical to pen them separately from the older calves.

A quarantine area for newly introduced calves could be specifically designed for thorough cleaning and disinfecting, with sufficient pens to rest each one for at least two weeks before occupancy with a new calf. Pressure hoses should be used for cleaning while a steam cleaner may reduce the need for disinfecting pens. There is less chance of disease build-up by using galvanised iron rather than wooden pens and fittings since dried faeces are easier to remove from iron fittings. Poorly cleaned wooden pens can provide an ideal medium for bacteria to grow.

Some calf rearers choose to run the calves together in pens or even outside on pasture then individually stall them for milk feeding. One operator even locks the calves up each night following milk feeding and gives them access to pasture, grain and hay throughout the day. One or two dogs help to quickly round up the calves each afternoon for milk feeding. By keeping the milk feeding area separate from the pens or pasture, it can be thoroughly cleaned each day.

Rather than wean calves onto pasture, some operators house them for several weeks to ensure effective rumen development. Ideally, weaned calves should remain in their milk feeding pens for a few days after weaning so they can get used to their new feeding regime before having to adjust to new pens and pen-mates. If the concentrate mix fed during milk rearing is different to that fed after weaning, both should be on offer (separately or mixed together) for the few days after weaning.

When group housing calves, the optimum group size is 10 calves or fewer, making it easier to regularly check the animals with one or two glances at each pen. Ideally, calves should not be moved from one group to another. Sick calves should be moved to hospital pens and, once recovered, to a new group. No dead calves should be replaced.

To restrict the likelihood of cross infection of diseases, the number of calves in any one 'air store' (shed or compartment of a shed) should be limited to 60. Ideally, the air store should be on an all-in/all-out basis. Cleaning, disinfecting and resting of these pens post-weaning are just as important for the health of weaned calves as they are for milk-fed ones. Hospital pens will obviously require more thorough cleaning to ensure removal of all residual pathogens.

Physical comfort of calves

The ideal temperature and humidity for calves is 17°C and 65% relative humidity. However, a normal healthy calf, eating well, is remarkably cold tolerant and is hardly affected by air temperatures below freezing point, provided it is dry and not exposed to draughts. On the other hand, sick calves, particularly emaciated ones with poor appetites, are very susceptible to cold. The hospital pens should be in a warmer part of the calf shed and be able to be heated if required.

Table 11.1 shows the temperatures at which young Friesian calves start to shiver. Equivalent temperatures for Jersey calves are 4–6°C above those for Friesians.

Table 11.1 Environmental temperatures at which young Friesian calves start to shiver

	Normal feeding	Low level feeding
Dry coat, no draught	3°C	12°C
Dry coat, draught	8°C	17°C
Wet coat, draught	13°C	19°C

It clearly shows the importance of level of feeding and protection from draughts on calf comfort. Calves lying on dry concrete lose more heat than those lying on wooden slats or damp straw. The warmest bedding is deep dry straw, wood chips or rice hulls. Draughts coming up through wooden slats or metal grating floors should be eliminated during winter. These can be easily detected using a lighted candle or match. On the other hand, draughts on hot summer days improve comfort by decreasing heat loads on shedded calves. Heat stress can also be reduced through constructing sheds with insulated roofs and well-ventilated walls, and by feeding calves in the cool of the evening.

Types of flooring

The floor of calf pens is the surface on which animals stand, walk, lie down and pass excreta. It must, depending on the needs, be solid, non-slippery and well drained, or comfortably soft, warm and dry, and easy to clean by machinery. No single material meets all these specifications. Of most importance, it must provide calves with a secure footing.

Wooden slats (50 mm x 25 mm with 20 mm gaps) placed 150–200 mm above a concrete sloping floor are ideal for calves. This arrangement allows the urine and dung to pass through the grating onto the concrete below where it can be easily removed by hosing without unduly wetting the calves. The grating should be made in sections small and light enough to be removed from the calf shed for thorough cleaning and disinfecting so as to prevent any build up of disease.

Wire mesh suitable for calf shed flooring is also available. Designed specifically for pigpen floors, it incorporates a mesh opening of 12.5 mm x 150 mm. It is welded onto a metal frame and should be positioned about 150–200 mm above the concrete floor. In one calf shed, metal floors have been placed directly over a septic pit.

A deep litter of rice hulls or wood chips 40–50 cm thick over a concrete floor is probably the best bedding material. With occasional topping up and removal of excess dung, rice hulls can stay clean and dry for up to four months in summer. In winter, they may need replacing every couple of months. Access should be provided for a tractor with a front-end loader or scraper to clean pens between batches. Tractor traffic must then be considered when planning concrete floor thickness. A damp-proof membrane should always be included in concrete floors.

Sheep shearing sheds are not ideal for rearing calves because the narrow gaps between wooden slats stop faeces easily falling through, while draughts can come from under the floor.

Alternative flooring could be concrete in the feeding area and a rice hull, wood chip or sawdust deep litter at the back of each pen. Dry straw is excellent but is very labour intensive. When given a choice, calves seem to prefer sawdust and rice hulls to straw and wooden slats, and they least prefer metal gratings. Despite these preferences, there is little difference in performance of calves raised on either straw bedding or wooden slats.

Effluent disposal from pens is most important. A minimum fall of one in 20 will ensure that free liquid drains away. Drainage channels should run under the feed and water buckets at the pen fronts and drain both pens and passages. The calf rearing unit at Kyabram Dairy Centre is designed with each pen draining through the back wall into a drain outside the shed while the passage is drained separately. All drains run into a sump with a manure pump (that also handles rice hulls) for disposal of pen effluent and rainwater.

Figure 11.1 shows half the shed in operation with rice hull bedding and the other half awaiting more rice hulls.

Figure 11.1 The Kyabram Dairy Centre calf rearing shed

Figure 11.2 Inside the Kyabram calf rearing shed

The floor area should allow for 1.5–2 m² per calf. It could be less for group pens than individual pens because calves can lie in any direction and so, in theory, no corners are wasted. An individual pen 1.5 m x 0.8 m is the minimum required up to 4 weeks of age, while a pen 1.8 m x 1 m will suffice for an 8-week-old calf. Pens can be converted from single to multiple by having removable pen divisions such as those shown in Figure 11.2.

With a 35 cm/calf pen frontage and a 1.5 m²/calf total area, group pens should be 3–4 m deep. In fully enclosed sheds, the stocking density is generally limited by the air volume per calf, rather than the floor space. The recommended cubic capacity is 6.5–7 m³ per calf with a ceiling height of at least 2.7 m.

Feeding and handling facilities

Feeding space requirements for individually fed calves (whether individually or group penned) is 35 cm/calf. This may limit the shape of group pens with bucket feeding to allow for sufficient frontage to the feeding passage. When feeding concentrates from a trough or bulk feed bin, allow 10 cm/calf. Troughs are more versatile but less protected from birds and vermin. Hoppers must be robust and provide an even flow of feed without blocking or bridging. They should also be kept clear of the built up manure. A 10-week-old calf is able to reach into a trough 70 cm high from the ground. If you are restricting or controlling concentrate intakes, 20–30 cm feeding space/calf is required.

Milk replacers can be mixed in stainless-steel tanks using warm water and with electrically powered rotors. It can be pumped or gravity fed to buckets using a petrol bowser dispenser. Feed scales are essential to ensure accurate weighing of powder. Whole milk can be pumped directly from the milking parlour to buckets or even to feeding drums located in nearby paddocks. If transition milk is being preserved, a milk line could be used to take it directly from the milking parlour to the preserving tank.

Portable milk tanks with delivery hoses could be used when rearing calves in hutches. Whatever the method, regular flushing of hoses with water and thorough cleaning of milk dispensers and buckets is important. Concentrates can be automatically handled and dispensed using large silos and conveyors of the auger, chain or endless belt type, such as those used for supplementing dairy cows during milking. Straw can be chopped then fed out in troughs or more easily handled unchopped using hayracks.

Water can be supplied through troughs, bowls or nipple drinkers with one communal trough per two pens, one water bowl for up to 20 calves or one nipple drinker per five calves. Calves can drink up to 15 L/day, and 25 L/day on hot summer days. Drinkers must be guarded against pollution by faeces (both from calves and birds) or damage by rubbing. Polythene piping may require protection from chewing, but, surprisingly at Kyabram, calves do not seem to chew the polythene piping leading to water bowls. Water bowls and troughs should be the low-pressure type, as cattle tend to play with water. They should be sited away from the lying area so that spillage may drain away freely.

Cattle handling facilities should incorporate a calf race, head bail and calf scales and lead to a loading ramp. A race for adult cattle can be modified for calves by adding a partition to make it narrower. This should be at least 1 m high and should reduce the width of the main race to 40–45 cm. In group pens, self-closing yokes at the feed face allow for easy restraining of calves for closer inspection or veterinary attention.

Passages between pens should be wide enough (1.2 m) for carrying buckets in both hands, be easy to clean and self-draining, and have non-slip surfaces. Calves should be able to be easily removed from their pens for weighing.

One very important item for any calf shed is a centrally located whiteboard plus erasable marker pens. Managers can list jobs for staff, staff can list details of calves requiring particular attention, and staff can leave instructions for mobile calf-sale operators. To assist in feeding and health management, every pen should be clearly numbered to remove ambiguity when recording feed intakes and calves requiring attention, etc.

Calf scales

Weighing scales are an essential component of good cattle handling equipment. They are important for monitoring the growth of calves during rearing and ensuring grazing management is sufficient for growing heifers to achieve target weights. Chest girths have been recommended in the past for estimating calf and heifer development, but they are not accurate enough. They can overestimate the weight of a 50 kg calf by up to 15 kg.

One example of the benefits of calf scales shows if bobby calves are sold for slaughter through calf scales. Payment is generally on the basis of cents/kg live weight and this increases with different categories of live weight. By regularly weighing bobby calves before sale, producers could ensure higher returns through selling calves in a heavier weight category. For example, if 39 kg calves sold for 80c/kg, they would return $31.20 each, whereas, if 41 kg calves sold for $1/kg, they would return $41 each. By holding onto calves for a little longer until they weighed 41 kg rather than 39 kg, producers would get $9.80 for that extra 2 kg live weight gain.

Beef producers require scales to plan feeding strategies for different production systems so their animals can meet end-point specifications. Cattle scales also allow dairy farmers to monitor changes in cow weights throughout lactation. This will ensure that cows are being fed and managed properly to take advantage of their ability to utilise body reserves for milk in early lactation then replace it later in lactation. Changes in body condition score are a guide to this, but weight changes are the

Figure 11.3 Calf scales are an important component of any rearing system

ultimate measurement. Scales can also be used to check weigh bales of hay or silage to assist in supplementary feeding programs. Ideally, calf scales and yards should not be used by mature cattle; this is particularly important if Johne's Disease is a problem.

Cattle scales are not costly. Clock face scales cost $1000 or less, while electronic cattle scales cost from $1000 to $2500 (Moran and Stockdale 1996). They are equipped with digital readouts and can store live weight data for downloading onto farm computers.

Cleaning and sanitising feeding equipment

Dairy farmers are now penalised for poor quality milk, so have had to become more aware of the principles of cleaning and sanitising their milking harvesting equipment. What about their milk feeding equipment in the calf shed? How often do they clean and sanitise it? Much of this equipment may, in fact, be stored in conditions for ideal bacterial growth, namely moist, no direct sunlight and with poor air exchange. Milk feeding out of dirty buckets and teats are common ways to spread scouring pathogens from one calf to another. Ideally, all feeding equipment should be cleaned and sterilised between feeds.

Cleaning removes residual milk from surfaces, while sanitising (or sterilising) removes bacteria from cleaned surfaces. The principles of good cleaning and sanitising can be summarised as WATCH, namely:

- Water: good water quality is important.
- Action: such as mechanical action with a scrubbing brush.
- Time: leave equipment long enough for the chemicals to work.
- Chemicals: match the chemicals for the job, detergents for cleaning and sanitisers for sterilising.
- Heat: chemical activity doubles every 10°C over 50°C.

It is best to start with a rinse using luke-warm water (as hot water can bond milk residues to plastic), then wash with hot soapy water (plus a brush) to remove all the milk residues from buckets, teats, mixing containers (for milk replacers) and storage equipment. Following a second hot water rinse, if desired, sanitising can take place. Simple household bleach may be all that is needed, but the water should be 75–80°C. Teats and bottles can be immersed in a 20 L bucket for as long as the water stays hot enough. Some farmers only sanitise the feeding equipment just prior to the next feeding, with the youngest calves fed first.

To complete the sanitising process, it is important to allow all equipment to completely dry before reuse. A drying rack will keep clean equipment off dirty floors, as well as improve air circulation.

Calf sheds and children

Unfortunately, with ever increasing stories about accidents and deaths on farm, many of which involve children, owners and managers of calf rearing operations must become more aware of the dangers for children in calf sheds. With increasing surveillance to comply with Occupational Health and Safety requirements, staff need to be protected from work-related accidents.

When purchasing chemicals for veterinary or cleaning/sanitising purposes (whether they be specifically for calf rearing or for other farm uses) ask for the associated Material Safety Data Sheets and store them in a secure but easily accessible place. These sheets will provide you and your doctor with information essential to treat any accidental spillages or swallowings, the latter more likely by children. Keep all chemicals out of reach of children, and preferably in a locked cabinet. Place a first-aid kit in the calf shed or a nearby office or staff room and make all staff aware of its presence.

It goes without saying that all children love calves, particularly those that live in towns or cities. When they visit the calf shed, keep an eye on them: either yourself or one of your staff. Even small calves can become unsettled and injure a small child if unable to move out of the way.

Milk-fed calves, being monogastrics, have many of the same diseases as humans. These are called zoonoses. The most important ones include:

- Salmonella, E. coli, cryptosporidia.

- Ringworm, mange and other skin diseases.
- Leptospirosis, which is more of a problem with adult cows. In Victoria alone there were 91 cases recorded in 1992, this reducing to 25 cases in 1999, while in Queensland there were 231 recorded cases in 1999.
- Q fever. In 1999, there were 26 recorded cases in Victoria, and more than 400 cases in Queensland. A recent NSW survey found 33% of all dairy farm workers were infected with Q fever.

In past years, doctors located in dairying regions of California have noted a close association between outbreaks of specific pathogens causing calf scours and its occurrence in very young children. Similar incidences have been reported in rural papers in Victoria. When children visit the calf shed, extra precautions should be taken with their personal hygiene. They should be made to carefully wash their hands and face prior to eating. Ideally they should change their footwear and even clothes. If they frequently visit the calf shed, for example, to assist with feeding the calves, they should have a pair of boots specifically for calf shed use. Elderly people, with a reduced immune system, could also be susceptible to zoonoses.

Summary

In intensive calf rearing, disease control should be through sound management principles rather than preventive medicine. The essential aspects of calf housing can be summarised as follows:

- Don't overcrowd the shed or make it too big.
- Keep to a strict program of introducing new calves and don't get carried away with more calves than planned.
- Have an all-in/all-out system together with adequate cleaning between batches.
- Keep feeding methods simple.
- Check for draughts and rain at calf level.
- Minimise stress on calves and calf rearer.
- Take precautions against zoonoses that can affect people, particularly with children.

References and further reading

Mitchell, D. (1981), *Calf Housing Handbook*, Scottish Farm Buildings Inv. Unit, Aberdeen.

Moran. J. and Stockdale, R. (1996), *Weighing and Condition Scoring of Replacement Heifers and Dairy Cows*, Agnote AG0505, Melbourne.

Tasmanian Department of Primary Industries (1991), *Rearing Dairy Replacements. A Manual for Dairy Farmers*, Dep. Prim. Ind., Hobart.

twelve 12

Welfare aspects of calf rearing

Calves born to dairy cows are routinely submitted to more insults to normal development than any other farm animal. They are taken from their mother generally within their first day of life, often not even allowed to suckle colostrum immediately after birth, and frequently deprived of their most natural feed, whole milk. They may be fed one of the various cheaper liquid substitutes for milk, but because these are still more expensive than solid foods, many are weaned off milk as quickly as possible.

Many calves are transported from their farms of origin into marketplaces and/or onto specialist rearing units. Such calves are not only denied the opportunity to feed normally from their mothers, but are also subjected to the rigours of travel and, if they pass through calf auctions, they can be exposed to a high risk of disease. This all occurs when at their most susceptible stage of life.

For up to a million calves every year, their lives are very short as they are slaughtered within one week of birth. Because they have little economic value, such calves can be subjected to practices, which contravene accepted codes of practice for their welfare. The chances of young calves surviving their first week of life depend on their potential value as replacement heifers or vealer mothers if female, or their potential for producing beef or veal if male.

Generally speaking, the concern for the calf's welfare improves with its potential value. This could be disputed when considering traditional European white veal systems but, in Australia, newly born bull calves slaughtered as bobby calves are less well looked after than those to be grown out for pink veal or dairy beef.

As soon as the calf is removed from its mother, it is placed in an artificial situation and it is up to producers to try the best they can to provide conditions at least as good as those which would be encountered naturally. Some welfare lobby groups would argue that this means being prepared to give the baby calf the same individual attention that it would get from its mother. This has to be reconciled with the demand on the calf rearer to look after large numbers of calves in an efficient and timely manner.

Calf housing is still sub-standard on many farms and far too many calves suffer from digestive and respiratory diseases. A calf spends much of its time lying down. Floors that are wet or too cold to lie on will increase heat loss. If calves are housed at or below temperatures that induce shivering, they will be using the energy supplied by the feed less efficiently. Maintenance of warm and dry bedding for calves depends on good drainage from under the bedding. If the floor is laid to correct slope and the drainage outlet carefully planned, free liquid is quickly led away from the calf and out of the building.

It is generally agreed that correct animal welfare equates to good animal performance. Stressed calves will not eat normally and grow efficiently.

Producers who ignore the basic fundamentals of animal welfare pay for it through higher feed costs, larger veterinary and drug bills, higher calf losses and, of equal importance, a poor image of dairying to the rest of the community. With heifer replacements, poor calf performance can carry through to low live weights at first calving, poor first lactation milk yields and even early culling due to infertility or low productivity.

Government codes of acceptable farming practice

Most western countries have recommended codes of practice for the care of calves, covering transport, housing, handling and feeding. Until recently, the best standards available were prepared by the Farm Animal Welfare Council of England and covered

Figure 12.1 Bobby calf welfare codes are currently under review

all domestic livestock. These stressed that animals should be 'provided with freedom' from hunger, heat and or cold stress, injury or disease and fear or stress. In conclusion they stated, 'the first essential of any good husbandry system was that it catered for the health and behavioural needs of the animals'. These should be the yardsticks for calf rearers to assess the quality of their management.

In 1992, the Federal Government published a code of practice for the welfare of cattle (SCA 1992), while in 1998, the Victorian Government published another (NRE 1998). The NRE code on the marketing and transport of bobby calve has already been presented in Chapter 5. The NRE code of acceptable farming practices for other management aspects of calf rearing is as follows:

Artificial rearing of calves for dairy replacements or beef or veal production

- Housing for artificially reared calves should be hygienic, with adequate ventilation, climate control and lighting. Flooring should be well drained with adequate dry lying space for each calf. Flooring and internal surfaces should not cause injury and should allow easy cleaning.
- Careful attention to group sizes, access to feed, bedding, milking shed location, ancillary accommodation, lighting, air inlets and outlets, handling facilities and stalls can alleviate problems of health, stress and aggression.
- Calves are social animals and seek the company of other calves. Individual penning of calves during early rearing (two–three weeks) may be preferable for disease prevention and management and developing a liquid feeding regime. Where individual penning of calves exceeds three weeks, careful consideration should be given to the social needs of these animals.
- Calves require at least 2 L fresh or preserved colostrum or an approved substitute within the first 12 hours following birth. Calves should continue to receive colostrum (or transition milk) for the first three days after birth. Thereafter, they should be fed at least daily on liquid milk, commercial milk replacer or transition milk, in sufficient quantities to provide essential requirements for maintenance and growth. High quality pasture, hay, pellets or straw should be available to calves from no later than 3 weeks of age to help in development of their digestive tracts and to ease the stress of weaning. Hygienic calf feeding practices, including thorough daily cleansing of all equipment (feeding units, lines, bottles, nipples, troughs, etc.) may be required to protect calf health and welfare and to prevent diarrhoea.
- Milk replacers based on skim milk should not be fed to calves under 3 weeks of age, unless they are in a properly balanced formulated mixture of protein, fat and vitamins. Milk replacers should be reconstituted according to manufacturers' instructions. Milk and milk replacers should not be fed in excess of body temperature (39°C).

- Calves should be weaned off milk, milk replacer or transition milk onto rations providing all essential requirements only when their ruminant digestive systems have developed sufficiently to enable them to maintain growth and wellbeing. Weaning of milk or milk replacer may be opportune time to introduce calves to group housing. The process of weaning can occur as early as 3 weeks of age.
- Restricted rations of the 'white veal' type, that is, iron deprived (lower than 20 ppm), which cause anaemia, are unacceptable.
- Calf rearing systems in which calves are individually and continually housed in pens or cribs, the available floor area for each calf must take into account the normal behaviour of calves. The floor area must be sufficient to enable each calf to freely turn around, stretch out and lie down comfortably. A floor area of at least 1.5 m^2 should be provided for each calf individually housed in pens or cribs. Pen height should be a minimum of 1 m with provision of additional height to allow for adequate ventilation space.
- Social interaction is an important calf welfare need. In systems using individual pen or crib housing, visual contact between calves must be facilitated. This can be by allowing uninterrupted visual contact between calves at the front of individual pens, and by restricting the height of solid partitions between calves to a maximum of 50 cm from the floor and permitting social interaction and full vision of other calves.
- Every effort should be made to ensure an adequate flow of ventilation to housed calves. Calves must be protected from rain, wind and extremes of temperature. In cold weather, feeds with a high energy value should be provided.
- Where large numbers of caves are reared, they should be grouped by age and size to reduce competition for food and to allow closer observation and management.

Management practices

- Restraint should be the minimum necessary to perform management procedures efficiently.
- Procedures and practices that cause pain should not be carried out if painless and practical methods of husbandry can be adopted to achieve the same result.
- Procedures and practices applied to cattle must be competently performed.
- Any injury, illness and distress should be promptly treated.
- In any situation, supervision should be by competent stock persons.

Castration

- Castration with burdizzo should be performed as young as possible. The burdizzo method of castration involves crushing the cord of the testes and, being a bloodless operation, reduces the risk of infection.

- There is also a European technique of partial castration, called hemi-castration, which is used on bulls between 1–5 months of age, and consists of removing part of the testes to prevent the production of spermatozoa, but without damaging its ability to produce male sex hormones. This makes the bull infertile but still able to benefit from the production of the sex hormones, thus improving its potential growth rate and feed efficiency when compared with steers. Trials in Australia have shown that these animals outperform steers, but they still require the more sophisticated management of intact bulls, such as better fencing.
- Castration with rubber rings should be, ideally, performed on calves up to 6 weeks of age and, where operations and management make this difficult, not beyond 12 weeks.
- Castration by knife without local or general anaesthetic should be confined to calves under 6–8 months of age. Bulls over 6–8 months should be castrated using appropriate anaesthetic. Castration of mature bulls should, preferably, be performed by a veterinarian using anaesthesia.

Tail docking

- Tail docking may be performed only when necessary for udder or herd health. Tail docking should only be undertaken on young female cattle, preferably under 6 months of age. Surgical removal of the tail should only be performed with the use of anaesthesia.
- A minimum length of tail should remain, sufficient to cover the vulva.

Dehorning

- To minimise injury to other cattle, all horned cattle should be dehorned as young as possible and at a suitable time to reduce fly worry. After dehorning, cattle should be inspected until healing has taken place, and any infected wounds treated.
- Inward growing horns likely to penetrate or contact facial features should be trimmed appropriately.
- Dehorning of cattle without local anaesthetic or analgesics should preferably be confined to animals under 6 months of age. Older animals may be 'tipped' (ends of horns removed without cutting into sensitive horn tissue) without anaesthetic in order to reduce their potential to cause injury.
- Dehorning by means of chemicals is not accepted for any class of cattle. The recommended methods for dehorning of calves are by heat cautery, scoop dehorners or gouging knife.

Health

- Appropriate preventative measures should be implemented for diseases that are common in a district or are likely to occur in the herd. A suitable vaccination,

internal and external parasite control plan should be devised and followed for each farm.
- Internal medications, such as vaccines and drenches, and external medications, such as dips and pour-on formulations, should be stored and given in strict accordance with the manufacturer's instructions and recommended methods of administration. Overdosing may harm cattle and underdosing may result in failure of the medication. Expiry dates and withholding periods should be strictly observed.
- Cows with cancer eye should be culled or treated as soon as possible after cancer is noticed. Cancers must not be allowed to progress untreated simply to permit the cow to complete raising a calf.

Humane destruction of cattle
- The preferred methods of euthanasia are overdose of anaesthetic under veterinary supervision or using gunshot or captive bolt pistol by the frontal method. The captive bolt pistol or firearm should be directed at the point of intersection of lines taken from the horn bud to the opposite eye.
- An animal stunned with a captive bolt pistol must be bled out by severing the major vessels of the neck as soon as it collapses to the ground. To avoid injury due to the animal's involuntary leg movements, the operator should stand behind the neck.
- Killing day-old calves may also be achieved by a heavy blow to the crown of the head to stun the calf prior to bleeding out. All other methods of killing are unacceptable except under extreme conditions in which common sense and genuine concern for animal and human welfare should prevail.

Additional management practices not included in the above code

Heifers may be born with more than four teats. They can interfere with milking and spoil the appearance of the udder. Extra teats can be snipped off before the calf is 2 weeks old. When removing, the extra teat should be gently pulled away from the udder and cut with a pair of clean, disinfected and sharp scissors. The angle of the cut should run head to tail, so the scar will blend in with the normal folds of the udder. Iodine should be applied to the wound, and maybe fly spray during warm weather. If the extra teat is close to the base of a normal teat, a veterinarian should perform the operation.

Permanent identification of heifers from an early age allows for their easy recognition within the dairy herd and is essential for record keeping. It is now compulsory to eartag all calves at birth with tags showing the property identification code. Ear tags are easy to attach and those consisting of one large tag made from soft unbreakable plastic are best. Others tend to tear out and break. The tag should be inserted between the veins towards the middle of the ear as abscesses can form by inserting tags into veins. If one does form, a medicated spray should be applied until it heals. Tags can also be placed around a back leg, but these can be hard to read in muddy conditions.

Some producers clip off parts around the edge of the ear using a code to identify the particular sections removed. Some breed societies have other systems of identification such as ear tattoos, metal ear tags or heifer photographs. Government herd recorders in Queensland use ear tattoos in their recording service, while it is a legislative requirement to identify all cattle in Tasmania older than 6 months with a registered ear mark or tattoo. The tattoo number should be placed in the clear space towards the top of the right ear after it has been thoroughly cleaned with a cloth saturated with methylated spirits or soapy water and then dried.

Brands can be either acid, caustic, hot iron or freeze types. Freeze branding is best because calves suffer little pain and hide damage is minimal. It is simple and quick, involving the use of dry ice and alcohol mixture. Freeze brands stand out on dark haired animals, while hair that regrows in lighter coloured animals, such as Jerseys, is usually a different colour. However, freeze branding on a white haired area should be avoided. They should be applied to calves weighing more than 200 kg, as brands on smaller calves will distort as they grow.

The European Union has several guidelines on calf welfare that may become incorporated into future Australian guidelines (Sue Hide, personal communication). These include:

- For calves kept in groups, the unobstructed space allowance available per calf shall be at least 1.5 m^2/calf up to 150 kg, 1.7 m^2/calf from 150–220 kg and 1.8 m^2/calf from 220 kg or more.
- No calf shall be confined to an individual pen after 8 weeks of age, unless a veterinarian certifies that its health or behaviour requires it to be isolated in order to receive treatment.
- Individual pens for calves, except for those for isolating sick animals, must not have solid walls, but perforated walls that allow calves to have direct visual and tactile contact.
- Calves less than 1 week of age are not to be transported or slaughtered. Some member states have specified 14 days as the minimum age for transportation of calves.
- Animals must be fed and watered every eight hours and are entitled to a 24-hour rest at the end of their journey.

Australian Veterinary Association's policy on calf welfare

The Australian Veterinary Association in May 1989 published its policy statement on the welfare of vealer calves. The Association defines a vealer calf as one reared for the purpose of slaughter for human consumption up to the age of 6 months. They are fed extra concentrates to facilitate better carcass weight, and are usually housed intensively. Many of their policies are covered in the guidelines presented above, but there are several others that are not specifically mentioned. These are as follows:

- The calves should be housed in well-ventilated and well-lit surroundings. An acceptable light intensity is 215 lux or natural daylight. The light/dark ratio should be ideally 50:50, although increasing the light periods will increase feed intake and growth rate in the calves.
- The optimal ambient temperature for housed calves is 20°C with an acceptable temperature range of 10–25°C. Ventilation depends upon the type of shed, but should be sufficient to maintain temperature and humidity, and remove build up of potentially toxic products such as methane, carbon dioxide, ammonia and airborne microorganisms.
- Ideally calves should be housed in groups rather than individually. They must be able to see, hear, smell and touch other calves and have relative freedom of movement. There is an essential requirement for veterinary supervision in the rearing of calves in group housing. Short-term individual housing of calves would be preferable when calves are first introduced into the vealer unit to minimise disease spread. A calf must have room to stand, lie down and adopt a comfortable sleeping position on a dry floor. The size of the pen to be used depends upon the weight of the individual calves in the group pen. These are summarised in Table 12.1

Table 12.1 Recommended guidelines for housing vealer calves in group pens

Live weight (kg)	Minimum pen floor (m^2)	Length of pen (m)	Feeder space (cm/head)
<60	2.0	1.1	30
60–100	2.2	1.8	30
100–150	2.4	1.8	35
150–200	2.5	2.0	40

- Calves must not be fed an iron deficient diet. Apart from causing anaemia in the calves, deficient diets are not necessary or desirable for good meat colour. Available evidence would indicate that a diet containing 30 ppm of iron in the dry matter provides sufficient iron to prevent anaemia, although the meat colour still remains pale.
- Shedded animals should receive fat-soluble vitamins (A, D and E) since they have no access to pasture or sunlight. For the wellbeing of the calf, suitable fibrous food should be provided to allow natural rumination to develop, especially after 4 weeks of age.

- There should be minimal mixing of calves of different age groups to prevent the spread of infections from the older calves to the younger calves or else the introduction of disease from the newly bought calves.
- Diarrhoea is a common complaint of housed calves, with shed design being a vital feature in predisposing to infection. Diarrhoea must be treated as soon as it occurs and with the appropriate treatment. Pneumonia is also common in calf groups undergoing stress. A competent stock person, capable of early diagnosis and treatment of disease, must supervise the operation. Due care must be taken with respect to antibiotic residues in the meat of treated calves.
- Calves should be slaughtered at a site located within as reasonable a distance from the production unit as possible.

Key issues identified by the Animal Welfare Centre

The Animal Welfare Centre was established in Melbourne in 1997, comprising of staff from several Victorian government institutes and universities. Its primary function is to coordinate research, teaching and training in the welfare of farm, laboratory, companion and captive animals in Australia (Michelle Edge, personal communication). It is currently developing a welfare audit for the dairy industry under the direction of an advisory group of dairy industry, government, veterinarian and private welfare specialists. The industry sectors being addressed are calves (including bobby calves), heifers and mating management, the milking herd, as well as cattle transport and slaughter.

The Centre has prioritised welfare issues using three criteria. These are:

- Community concern, now or in the future if awareness is likely to develop.
- Welfare risks, such as stress, pain, injury and behavioural problems.
- Number of animals affected within Australia.

The process initially involved scoring each of these criteria, but this has been modified to categorising them as extremely important, very important or important. Those related to calf welfare are listed in Table 12.2.

Table 12.2 Welfare issues relating to calves as prioritised by the Animal Welfare Centre. Categories are extremely important (1), very important (2) and important (3)

Issue	Category
Transport and handling of bobby calves	1
Age of bobby calves for transport	1
Tail docking	2
Inspection and management just prior to slaughter	2
Stockmanship (handling, knowledge, motivation)	2

Issue	Category
Prompt effective treatment or euthanasia of sick, lame or injured animals	2
Housing and husbandry of bobby calves	2
Horn bud removal	2
Dehorning	3

Clearly, improving the management of bobby calves is considered the highest priority. However, the Centre has identified many other calf management practices, such as tail docking, removing horn buds and prompt treatment of sick calves as key areas to address.

Figure 12.2 Good dairy farmers pay close attention to calf welfare

Public lobby groups

Anyone rearing vealer calves must come to terms with the fact that they are being reared for slaughter. In fact, all cattle at the end of their productive life are eventually slaughtered. In our society there are groups of people, especially amongst the urban 'fringe' of the animal welfare lobby groups, trying to outlaw the slaughter of any livestock. In America and Europe they often target the more emotive slaughter of young calves, in particular, the white veal trade.

As well as these extremists, there are others who are genuinely concerned with the health and welfare of farm animals. In recent years the white veal industry has responded to both types of groups and this has partly lead to the principles of growing calves for pink veal through modifying diets and calf housing.

In the future, welfare lobby groups will obtain greater access to farms. Therefore, producers rearing calves as heifer replacements or vealer mothers or for veal or dairy beef will have to ensure that their levels of housing and management comply with society's standards. We now have excellent sets of welfare guidelines on which to base calf rearing systems (SCA 1992, NRE 1998).

It should be remembered that they are guidelines and not legislature. However, they will probably form the basis of future judicial decisions on what does and does not constitute acceptable calf rearing welfare practices.

References and further reading

Natural Resources and Environment (1998), *Code of Accepted Farming Practice for the Welfare of Cattle*, Bureau of Animal Welfare, Agnote AG 0009, Melbourne.

Standing Committee on Agriculture, Animal Health Committee (1992), *Australian Model Code of Practice for the Welfare of Animals*, Cattle, SCA Rep. Series No.39, CSIRO, Melbourne.

thirteen | 13

Post-weaning management

This chapter discusses the post-weaning management of calves reared as replacement heifers for dairy farmers, as well as systems for producing beef from dairy calves. It summarises the major points of my recently published book on heifer rearing (Moran and McLean 2001, see References and further reading).

On-farm rearing of replacement dairy heifers

All too often, dairy farmers do a good job of rearing heifer calves up to weaning but then virtually neglect them thereafter.

Weaned growing heifers require less attention than milk-fed calves and milking cows. From weaning until breeding, and sometimes even after then, daily contact is not necessary. Because their nutrient requirements are relatively low compared to lactating cows, many heifers are located away from the prime grazing areas on the dairy farm, often on 'run-off' blocks or on agistment. Unfortunately, the saying 'out of sight, out of mind' applies too frequently to replacement heifers. This relative neglect is understandable in view of the long time it takes any inadequacies in post-weaning practices to be reflected in poor milk production.

Dairy heifers need to be well fed between weaning and first calving. Growth rates should be maintained, otherwise heifers will not reach their target live weights for mating and first calving. Undersized heifers have more calving difficulties, produce less milk and have greater difficulty getting back into calf during their first lactation. When lactating, they compete poorly with older cows for feed and because they are still growing and will use feed for growth rather than for producing milk. They are more likely to be culled for poor milk yield and/or infertility.

The onset of puberty is related to weight rather than age. A delay in puberty could mean a later conception, which can disrupt future calving patterns. All heifers should reach a minimum weight before joining, as lighter heifers have lower conception rates. Target live weights at mating and first calving are discussed below.

To recommend detailed post-weaning management procedures for dairy heifers is not practicable. The system adopted should reflect local and climatic conditions and personal preferences. The extremes of weather and availability of grazed and purchased feeds are probably the most important variables. Although replacement heifers are essentially non-productive animals, some expenditure is necessary. They represent capital and investment in the dairy herd's future. Heifer rearing should achieve the maximum return on this investment with a minimum of outlay. It should not be regarded as a haphazard undertaking, which hopefully will produce a pregnant heifer, but rather as a business enterprise with clearly defined goals such as:

- The number of animals to be reared.
- Their desired age at first calving.
- Their target live weight at calving.
- Their feeding program.
- Any housing and health requirements.

When rearing dairy replacement heifers, producers should have five major objectives:

1. The maintenance or expansion of herd size. Heifer rearing systems should provide sufficient animals to replace cows culled from the milking herd and allow for increases in herd numbers if required.

2. Calving by 24 months of age. Entry into first lactation by 24 months of age minimises the total non-productive days and maximises lifetime productivity.

3. Sufficient growth for minimal dystocia at first calving. Heifers need to be large enough to calve without difficulty.

4. Maintenance of health. The prevention of clinical and subclinical disease plays a large role in the ability of replacement heifers to meet live weight and age targets at first calving. Longevity and lifetime productivity is also affected.

5. Genetic progress. Replacement heifers generally have higher genetic merit than the current milking herd. This can be expressed as increased productivity (both milk volume and solids), improved efficiency of production and/or enhanced resistance to disease.

When considering these objectives, producers should decide whether to rear their own replacements on-farm, to have them contract reared off-farm or to purchase in-calf heifers. The two latter alternatives will save land for milking cattle, which is important where land is the major constraint to production, but are likely to cost more than on-farm rearing. When purchasing in-calf heifers, there is no guarantee that their genetic merit is superior to that of older cows in the herd and their health status is largely unknown.

Heifer rearing is not cheap, costing $800–$1000 to put a lactating first-calf heifer on the ground. This can account for 15–20% of the total milk production costs. It is not good economics to try to cut back on heifer rearing costs, as lifetime profits will be reduced. For example, $300 savings prior to first calving can reduce total lifetime returns by $780 (Moran and McLean 2001).

The number of first calving heifers each year will depend on the replacement rate within the milking herd. This is the sum of the wastage rate caused by infertility, mastitis, low milk yield, old age, accidents, etc., together with the particular culling policy for that herd, whether this is to improve milk yield, feed efficiency or calving interval. The number of heifer replacements to be reared also depends on mortality rates during rearing, conception rates at first mating and the proportion of heifers reaching target weight for ages.

Recent US surveys indicate that from every 100 cows that calve, 93 calves will be born alive; half of them heifers. About eight of these 47 heifers will die before point of calving, while five are culled for inferior genetics and another five for poor performance, reproduction or health problems. This leaves only 27 as replacement heifers. Such data needs to be verified for Australia.

With 20–30 heifers per 100 cows introduced into the milking herd annually, at least 80% of the milking cows should be artificially inseminated to obtain that number of replacements each year. When determining the total number of calves to rear, consideration could be given to rearing additional heifers for sale to other dairy farmers and/or bull calves for dairy beef.

In seasonal calving herds, a consistent calving program is important and, to achieve this, heifers should:

- Reach puberty at about 12 months of age.
- Become pregnant at 14 or 15 months of age.
- Calve at 24 months of age.
- Return to oestrus and be mated within 70–80 days of calving.

In year-round calving herds, it is possible to calve heifers down at 20–22 months of age, but to achieve this, management and feed inputs must be high. Aiming for 24–27 months at first calving is a more realistic target, rather than at 36 months, which is all too common in Australia. Earlier first calving ages are easier to achieve with the smaller, more rapidly maturing dairy breeds such as Jerseys or Ayrshires.

Benefits of heavier heifers

Provided heifers are at least 18 months old, the younger the heifers calve, the higher their first lactation and mature milk yields, the more calves they produced and the longer their productive life in the herd. An additional benefit is a more rapid generation interval and, hence, a faster rate of genetic progress in the milking herd. Lifetime productivity reaches a peak in heifers calving at 25–27 months of age. Heifers calving at 24–27 months of age can produce 21,000 L milk over a seven-year productive life, compared to 18,750 L if calving at 30–33 months, and only 17,000 L if calving at 36–42 months of age.

Several studies have documented the long-term benefits from heavier calving weights in Friesian heifers. For every additional kilogram at first calving, heavier heifers produce 7 L extra milk in each of their first three lactations. Therefore, if heifers calved at 500 kg compared to 450 kg, they would produce an extra 350 L milk/lactation or 1050 L extra milk over their first three lactations.

As part of an Australia-wide survey of dairy herd fertility, involving over 33,000 cows, live weights of 2000 Friesian heifers from 69 seasonal calving herds in Victoria and Tasmania were recorded just prior to their first calving (John Morton, personal communication). Heavier heifers calved earlier and conceived more readily during their first lactation (see Table 13.1), indicating a decreased need to induce (or even cull) second calving cows.

Table 13.1 The effect of live weight at first calving (LWFC) on percentage of heifers calving in their first three weeks of the calving period, subsequent three week submission rate, six week in-calf rate and the proportion of heifers conceived between seven and 21 weeks during their first lactation

LWFC (kg)	Calved in first 3 weeks (%)	3 week submission rate* (%)	6 week in-calf rate* (%)	Heifers conceived from 7–21 weeks of mating* (%)
<400	36	58	49	30
400–440	49	74	60	27
440–480	55	77	68	21
480–510	65	82	68	19
510–540	53	85	75	13
>540	68	88	77	10

* Data are expressed as per cent of first-calf heifers mated.

Heavier calving live weights also reduce the incidence of calving difficulties and wastage rates, which both adversely affect lifetime performance and herd profits. The percentage of replacement heifers that either die or are culled before their second calving has been recorded at 30–35% in two Victorian studies, considerably higher than the target 20% possible with better-grown heifers.

Target live weights for growing heifers

Heifer milk production depends on their live weight at calving and how well they are fed and managed as milkers. Their optimum live weight at first calving depends on the milk yield farmers wish them to achieve at maturity in the herd. Table 13.2 presents data on target live weights (in-calf) for 2-year-old Friesians heifers required to produce a subsequent full lactation yield as mature cows. On most Australian farms, 6000 L milk/lactation would be a realistic target, meaning that heifers should be grown out to 500–550 kg at 2-year-olds just prior to calving.

Table 13.2 Target live weights for 2-year-old Friesian heifers to enable them to produce a specified milk yield as mature cows

Full lactation milk yield as mature cows (L)	Target live weight as 2-year-old heifers (kg)
3000	430
6000	540
9000	590

The first lactation yield of heifers can be a useful guide as to how well they are grown up to point of calving. Although their *absolute* milk yields can vary enormously with feeding management while milking, their milk yields *relative* to those of their herd mates is a useful criterion of heifer management. This value is determined by comparing the average full lactation milk yield of first lactation heifers with the average of the mature cows in the herd. Over the last 30 years, this value has increased in Australian herd tested herds from 65–70% to 80–85%. Friesian heifers have been grown out to produce 90% of their mature herd mates (producing 10,000 L/lactation) on Israeli feedlot farms. If this value is 80% or less, heifer rearing practices should be reviewed to establish if they are contributing to poor heifer production.

There is a critical period for the developing udder during which time excessive growth rates can increase the deposition of fatty tissue in the udder and reduce lifetime productivity. Exactly when this critical period occurs and exactly what constitute excessive growth rates have yet to be clearly defined, although there are some general guidelines. Live weight gains should not exceed 0.8 kg/day between 6 and 12 months. Heifers should not be fully fed during their second six-month period. This is unlikely to be a problem in pasture-grown heifers particularly with spring calving herds as it coincides with autumn and winter, a period of traditional pasture shortage.

A dairy cow will attain her mature live weight in about the fourth lactation and the object of the heifer rearer is to produce an animal of 75–80% of that weight by first calving. Target weights for dairy heifers should be easily achievable on well-managed dairy farms. Unfortunately, some producers use these target weights as the average rather than minimum, meaning that many of their heifers are below recommended live weights.

Traditional target weights are too low to ensure first lactation heifers attain their potential productivity, particularly on farms where milking cows are well fed. Table 13.3 summarises revised target weights for Jersey and Friesian heifers at various ages. The weights for Friesians more closely match the US guidelines, which seems logical because Australian Friesians are now genetically close to the US Holsteins.

Table 13.3 Target live weight ranges (kg) at various ages for well-managed Friesian and Jersey heifers

Age (months)	Friesian	Jersey
3 (fully weaned)	90–110	65–85
6	150–175	110–130
9	210–235	155–180
12 (yearling)	270–300	200–230
15 (mating)	330–360	245–275
18	390–420	290–320
21	455–485	335–365
24 (Pre–calving)	520–550	380–410

Target chest girths and wither heights are often presented as an aid to producers without cattle scales. However, most weigh tapes overestimate live weights in growing heifers. Cattle scales are not expensive and can serve many roles in farm management (see Chapter 11).

Feeding heifers to achieve target live weights

Before planning feeding strategies for growing heifers, it is important to set realistic target live weight for different ages. For Friesians weighing 100 kg at 3 months to reach a target of 550 kg at calving as 2-year-olds, they need to grow at 0.7 kg/day, compared to 0.5 kg/day if calving at 450 kg. Average weight for ages and the dry matter (DM) intakes of good quality pasture (containing 10–11 MJ/kg DM of energy) required to achieve this are presented in Table 13.4.

Table 13.4 Average weight for ages and requirements for energy and pasture DM in heifers grown out to 450 or 550 kg at 2 years of age

Age range (month)	450 kg at 2 years old			550 kg at 2 years old		
	Live weight (kg)	Energy intake (MJ/day)	DM intake (kg/day)	Live weight (kg)	Energy intake (MJ/day)	DM intake (kg/day)
3–6	125	33	3.0	132	38	3.4
6–9	175	41	3.8	196	49	4.6
9–12	225	49	4.7	260	60	5.8
12–15	275	57	5.6	324	71	7.2
15–18	325	65	6.6	388	82	8.6
18–21	375	78	7.6	452	98	10.4
21–24	425	110	11.6	516	133	13.7

Heifers require a high quality diet to grow at 0.7 kg/day. Table 13.5 presents the energy, protein, calcium and phosphorus concentrations of their diets to promote this rate of live weight gain. The limited rumen capacity of 3- to 6-month-old heifers means that they should be fed a ration containing as high an energy and protein concentration as that of milking cows.

Table 13.5 Dietary quality for heifers of different ages to grow at 0.7 kg/day

	3–6 months	6–12 months	>12 months
Energy (MJ/kg DM)	10.90	10.30	9.50
Crude protein (%)	16.00	12.00	12.00
Calcium (%)	0.52	0.41	0.29
Phosphorus (%)	0.31	0.30	0.23

Grazing management should allow for continuous heifer growth throughout the first two years. Uniform growth is not necessary and may be impractical with fluctuating pasture availability. However, heifers should never lose weight or grow slowly for long periods during their first year, as they may not achieve their ultimate frame size and/or mating live weight by 15 months of age. Yearling heifers can show some compensatory gain in their second spring following feed shortages the preceding winter. Recommendations for grazing and feeding systems will vary with different regions. Rather than depend on 'recipes', farmers should use target growth rates to plan optimum feeding strategies. To achieve 550 kg by 2 years of age, seasonal target growth rates can vary from, say, 0.5–1 kg/day. The two most difficult periods to ensure acceptable growth in spring-born heifers are immediately after weaning and during their first winter.

When heifers graze with older cows, they increase their chances of picking up infections and, hence, developing immunities to any diseases carried by the cows. These immunities can then be transferred to newborn calves via the heifers' colostrum. Heifers reared in complete isolation from cows are likely to become infected as they calve and come in contact with the milking herd for the first time. This coincides with the time when they should be in peak health to produce milk, get back in calf early and also overcome any stresses associated with their radical change in management. Previous to this time in their life, heifers were non-lactating animals, continually at pasture, whereas now they have become lactating animals with regular human contact twice each day.

If the milking herd has a history of Johne's disease or if there is a high chance that Johne's carrier cows have been introduced to the herd, then grazing options for young heifers are reduced. This and other aspects of disease are discussed in Chapter 10.

Heifers may be set stocked separately from other stock, they can strip graze ahead of milking cows or can be rotationally grazed behind the milking herd to clean up the paddocks. Options for grazing heifers are detailed in region-specific booklets such as those written for Queensland (by Wishart 1983), Victoria (by Donohue and others 1984), New Zealand (by Holmes and Wilson 1984) and Tasmania (by Tasmanian DPI 1991).

Feeding systems for heifers in Queensland specifically recommend continuous supplementation of energy and minerals to overcome these deficiencies in the tropical pastures. Grazing systems in other regions should (but they do not always) allow for strategic concentrate feeding when grazed pastures cannot fully satisfy the nutrient requirements of heifers growing at up to 0.6 or 0.8 kg/day. This can be particularly important with spring-born heifers that are pregnant during their second winter because body condition as well as live weight at calving will influence first lactation milk yields and fertility. The nutritive value of pasture and supplements are discussed in Chapter 8.

Agisting young stock off the farm has much to commend it as it allows dairy farmers to use all available feed supplies to produce milk while still having control over the disease status of the heifers and the genetic progress in the herd. However, farmers must be well aware of the supply and quality of pasture for their agisted stock, the responsibility for stock health while away from the farm and the security of the agistment area against theft and straying heifers as well as neighbouring bulls. The proximity of the area and its cost are probably the major factors that need to be taken into account. Agistment works well provided that it is cost effective and heifer growth is monitored to ensure target weights are achieved. The costs and benefits of contract rearing have been discussed in my heifer book (Moran and McLean 2001).

Young stock should be handled frequently. When entering the milking herd, they must find their place in the social structure and this may take less time if they are used to human contact. For example, they should be run quietly through the milking shed a few times before calving to start to settle into the milking routine. Grazing the heifers during their last months of pregnancy with the main herd of dry cows can accustom them to the competitive conditions with which they will have to cope during lactation. Hand feeding heifers for a few weeks before calving will provide extra feed to build up body condition as well as get them used to being handled. Other aspects of heifer management are summarised in an NRE Agnote (Moran 1997).

Using dairy stock for beef production

Australia is unique as being the only country in the world that consistently slaughters virtually all its week-old bobby calves, those bull and heifer calves excess to milking herd requirements. Annual slaughterings of three quarters to one million week-old calves have been a feature of our dairy industry for the last 30 years. Most other major beef producing countries have integrated their dairy and beef industries, such that beef is a major by-product from milking cows.

Although dairy cows need to produce a calf each year to continue milking, dairy farmers only require 25–30% of these calves to replace those cows that die or are culled from the milking herd, since the average productive life span for milking cows in Australia is four–five lactations. This can be as low as two–three lactations in some countries practicing intensive dairy feedlotting, such as the US and Israel, meaning that replacement rates can be up to 50%, or virtually all the heifer calves born. Therefore, from a national milking herd of two million cows, that leaves a million or more calves not required as dairy cow replacements. Up to 100,000 are reared each year for meat production in various forms, whereas the remainder are slaughtered as week-old calves, producing 25–30 kg carcasses mostly destined for low-grade manufacturing meat.

One must surely ask the question: 'Why is it more profitable in Australia to convert these excess dairy calves into low-grade veal rather than to grow them out for meat production?' The reasons are clearly related to low profitability margins from dairy beef.

Dairy beef production in Australia

A limited number of Friesian steers have been grown out and grass finished in southern Australia for at least the last 50 years. However, our initial importations of early maturing British beef breeds last century have created the perception that beef animals must have well-developed hind quarters and, until recently, substantial subcutaneous fat cover. The more recent commercial interest in large European beef breeds (in the 1960s) may have changed our thinking about using subcutaneous fat to indicate the level of intramuscular (marbling) fat in the carcass, but we still like our beef cattle 'blocky'. As carcass yield is related to muscling, poorly muscled dairy-type animals often have low carcass yields. Consequently, traditional beef producers have always considered (and many still do) dairy animals to be inferior meat producers than beef cattle.

Friesians, and to a small extent Jerseys, and their beef crosses have been used for a wide variety of beef systems in the past. Culled milking cows have always been slaughtered for manufacturing beef. White veal, from calves fed entirely on milk, was once produced for a small group of affluent consumers, although this is now virtually all supplied from week-old bobby calves. As with many European countries, welfare guidelines have outlawed the production of traditional white veal in Australia. Dairy crossbred stock play an important role as dams in vealer units, to ensure calves are supplied with sufficient milk until slaughter at 9 months of age. In fact, beef x Friesian dams are often superior to straight beef dams.

Until the early 1990s, small tonnages of Friesian bull beef were exported for manufacturing purposes and there is once again renewed interest in bull beef in southern Australia, but this time as quality prime beef. Commercial interest in pink veal in the 1980s was for 70 kg carcasses for domestic consumption, while in 1995 several trial container loads of 150 kg carcasses were exported to Europe. Growing out dairy beef feeder steers for grain finishing for Japanese markets showed potential for value adding dairy calves in the early 1990s. The major limitation of this particular market segment is that suppliers of Friesian feeder steers have consistently been offered lower prices for feedlot cattle than those supplying other beef breeds. Another 'new' dairy beef industry is bull beef, where bobby calves are reared on one farm then grown out and finished for slaughter on another. Current beef markets are for purebred Friesian or Wagyu x Friesian bulls.

Such a stop/start history of dairy beef in Australia has always made it virtually impossible for long-term planning by potential dairy beef producers.

Limitations to dairy beef in Australia

Several suggestions have been put forward as to reasons for such a chequered history of dairy beef in Australia. They generally come down to poor returns for finished stock, although fluctuating grain prices have also contributed. End users (retailers and exporters) can apparently obtain equivalent quality beef for the same or lower cost than dairy beef. Therefore, without a price or quality advantage, what is the future for dairy beef?

Dairy cattle have been bred to produce depots of body fat that can be rapidly deposited and utilised as energy sources for milk production. Until quite recently, there has been little emphasis on them as dual-purpose animals, meaning that their carcass attributes (such as body conformation, subcutaneous fat and meat tenderness) have not been included in breeding programs. Friesians do not produce blocky carcasses with high levels of subcutaneous fat and, unless they are intensively finished, will not marble as readily as early maturing beef breeds. Jerseys produce yellow fat, which detracts from their marketability, even though their meat is often succulent and tasty. As such genotypes are not ideally suited to traditional beef markets, this limits their potential within the beef industry. The niche markets of 4- to 6-month-old veal or 24- to 30-month-old heavily marbled beef have never been features of Australian consumer requirements

The high cost of calf rearing

Artificial rearing of replacement heifer calves is a specialist job that dairy farmers generally have to undertake. As it is the most expensive period in an animal's life, calves reared for dairy beef would need to be grown out for slaughter at much older ages to dilute these high feeding costs. Unless these finished animals realise reasonable returns, dairy beef is unlikely to be profitable.

No matter what the system of beef production, the maintenance of breeding stock and the production of their offspring to replace those slaughtered for human consumption is a major part of the total feed inputs. With beef producers, these must be included in their total production costs. However, with dairy beef producers, these are 'paid for' by dairy farmers. As dairy calves are by-products of the milk industry, dairy beef farmers should then have lower production costs than beef farmers. Their ultimate lower carcass returns can allow for this, but they have rarely been sufficient for dairy beef to have a long-term viable future.

Producing feeder steers

Until producers of dairy feeder steers can be assured of long-term contracts for their stock, it is difficult for such an industry to have an assured future. The price of cereal

Figure 13.1 Purebred Friesian steers make excellent feeder cattle for feedlot finishing

grain will always be a major factor in the beef feedlotting industry. However, there are other feedstuffs, such as high-energy silages, that could be incorporated into beef feedlot rations, thus reducing their reliance on world stocks of cereal grain. Nevertheless, very few feedlotters appear to be genuinely interested in Friesian feeder steers, despite their obvious meat producing abilities in feedlots and their apparent demand by Japanese meat processors. Some optimists believe that the demand for Friesian feeder steers will be maintained and even increase in the future. However, most potential dairy beef producers seek further proof of genuine industry interest.

Producing pink veal

The small domestic pink veal industry that evolved in the 1980s foundered because consumers were not prepared to pay enough for consistently high quality veal. Quality control is sometimes lacking in new industries and pink veal in the 1980s was no exception. When producers require high carcass returns (at least $5/kg carcass weight, Moran and others 1991), there are no short cuts in their production systems. Pink veal animals must be housed and fed diets to grow at more than 1.2 kg/day during the final stages of finishing. Further details are summarised on an NRE Agnote (Moran 1996).

As with any new farming venture, it is important to find reliable markets for the end product before embarking. This is very important with high cost ventures such as pink veal because without the guarantee of high carcass returns, profit margins can be very small, even negative.

Figure 13.2 4-month-old pink veal steers

The 'new' bull beef industry

This market was initially developed in the mid 1990s by an abattoir in Warrnambool. The project required purebred Friesian bull calves to be artificially reared to reach 105–120 kg by 12 weeks of age, after which they were grown out on pasture to 550–600 kg. The bulls can either be backgrounded on one farm then finished on another, or the complete grass feeding stage could be undertaken on the one farm. Well-managed Friesian bulls should achieve slaughter live weights by 16–18 months, whereas British-breed steers generally take two years or more. Innovative grazing systems have been devised for the grass-finishing phase, involving strip grazing at high stocking rates with minimal supplementation.

The Wagyu x Friesian bulls are destined for Japan, a premium market requiring marbled beef not acceptable to most westerners. The company developing this market provides technical support in calf rearing, cattle management, animal health, pasture production and grazing management, as well as professional accounting and information technology.

References and further reading

Donohue, G., Stewart, J. and Hill, J. (1984), *Calf Rearing Systems*, Vic. Dep. Agric., Melbourne.

Holmes, C. and Wilson, G. (1984), *Milk Production from Pasture*, Butterworths, Wellington, NZ.

Moran, J. (1996), *Producing Pink Veal from Dairy Bull Calves*, Agnote 0567.

Moran, J. (1997), *Health and Mating Management of Heifers from Weaning to First Calving*, Agnote 0506.

Moran, J., Hopkins, A. and Warner, R. (1991), 'The Production of Pink Veal from Dairy Calves in Australia', *Outlook on Agriculture*, 20, 183.

Moran, J. and McLean, D. (2001), *Heifer Rearing. A Guide to Rearing Dairy Replacement Heifers in Australia*, Bolwarrah Press, Victoria.

Tasmanian Department of Primary Industries. (1991), *Rearing Dairy Replacements. A Manual for Dairy Farmers*, Dep. Prim. Ind., Hobart.

Wishart, L. (1983), *The Dairy Calf in Queensland*, Qld. Dep. Prim. Ind., Brisbane.

fourteen 14

Economics of calf rearing

Calf rearing is not a cheap enterprise and producers who try and cut costs will eventually pay for it in the long run.

Because of the uncertainties of Australia's weather, there will always be unusually cold, wet or hot spells, and calves are very susceptible to these changes unless provided with some form of housing. Inefficient milk feeding and cleaning systems require more labour, and despite what many producers believe, labour is not free and not even cheap. 'Cut price' milk replacers are generally cheap because they are lower in quality than normally priced powders, often because of poor processing techniques.

Calves can be reared on less whole milk or milk replacer than is often fed, provided their feeding and management allows for early rumen development. Once calves are weaned, poor feeding practices such as grazing low quality pastures or being hand fed low quality roughages, together with inappropriate concentrate feeding regimes will lead to slow growth. If dairy heifer replacements do not achieve realistic target live weights, long-term milk yields, reproductive performance and longevity will suffer. Sub-optimal growth rates in animals grown for dairy beef increase slaughter ages and can adversely affect carcass and meat quality. Money spent on good rearing and growing out practices will be recouped in improved returns for milk or meat.

Whole milk should not be put into the bulk milk vat for periods of up to eight days after calving; recommendations on this vary in different states. During this period cows produce colostrum and transition milk, which should all be fed to calves. If you are only rearing heifer replacements for the dairy herd, the colostrum produced by cows calving bull and cull heifer calves should provide ample liquid feed for milk-fed calves. Assuming a 25% replacement rate in the dairy herd together with 45 L available from each cow to rear heifer replacements, this provides 180 L of transition milk for each heifer calf. This is sufficient to rear a calf from birth to weaning. There should be little need for dairy farmers to buy milk replacers or use marketable whole milk to rear their calves.

Costing different feeds for calf rearing

Various methods for costing whole milk and milk replacers have been described in Chapter 7. Costs can be expressed in terms of either cents per kilogram (c/kg) of dry matter (DM) or cents per MJ of metabolisable energy (ME) in the product. The latter is calculated from fat and protein levels in the whole milk or milk replacer. For comparative purposes, in Table 14.1, whole milk has been assumed to contain 4% fat and 3% protein, while milk replacer powder has been assumed to contain 20% fat and 25% protein.

Costs for solid feeds such as concentrates or roughages can be calculated in a similar manner to liquid feeds once their cost in dollars per tonne and their DM and ME contents are known. Costs for purchased feeds are easy to calculate, but costs for home-grown feeds are more difficult to determine. Many economists use the opportunity cost of the feed as the basis of its pricing. This is the value of that particular feed if it was sold on the open market. For example, wheat can be grown on-farm for, say, $150/t, yet could be sold for $200/t. It should then be priced at $200/t because that is its actual value to the grower.

Chapter 8 presents several tables of DM, ME and protein values of selected feeds to provide guidelines on their nutritive values, however, these can vary considerably for any one feed type. It is strongly recommended when formulating rations for weaned calves, or any livestock for that matter, that actual measures of DM, ME and protein be obtained from commercial feed evaluation laboratories.

Energy and protein-rich feeds can be purchased ready-mixed and pelleted as commercial pellets or they can be blended on-farm from the raw ingredients to form a formulated concentrate mix. Calf rearing pellets often contain vitamin and mineral additives. Commercial pellets, despite being more expensive than on-farm mixtures, are usually the preferred solid feed for calf rearers. For the comparison below, pellets have been priced at $350/t and on-farm concentrate mixes at $250/t. This would be the price for an on-farm mix consisting of 80% rolled wheat (at $200/t) and 20% cottonseed meal (at $350/t) plus $20/t for blending and handling. These prices are for the concentrates delivered to the farm, which contain only 90% of their weight as dry matter – this should be taken into account when comparing the cost of different feeds.

Conserved pasture hay or silage can be priced on its opportunity costs in the open market. However, grazed pasture cannot be priced this way because it has less value as standing feed than when grazed and utilised by calves. Many farmers undervalue the cost of grazed pasture on their farm. After including the actual cash costs (such as fertiliser, weed control and irrigation), the indirect costs (such as fencing, repairs and maintenance on farm machinery), as well as the costs for labour and depreciation of farm machinery, grazed pasture is not cheap. Some economists even include council or shire rates and the return on total capital invested in land and equipment when calculating the real cost of pasture. Finally, these are all costs for producing the pasture in the paddock, but it must be remembered that about half of that grown is actually eaten by grazing animals. Taking all of these factors into account, grazed pasture can cost more than $120/t DM, which may not be all that cheaper than producing a grain or a forage crop. However, for this comparison, pasture costs will only include actual and indirect cash costs using a value of $60/t DM for the grazed pasture.

The relative costs of various feeds used in calf rearing using assumed DM, ME and protein values for 'typical' feeds of each type are presented in Table 14.1. In terms of energy, whole milk is generally cheaper than milk replacer, but both are two to three times more expensive than the energy supplied by concentrates. The cheapest source of feed energy is grazed pasture, which costs 17% for the same amount of energy contained in concentrates and only 5% that in liquid feeds.

Table 14.1 Costs for dry matter and energy in various calf feeds

Feed	Dry matter (%)	Energy (MJ/kg DM)	Protein (% DM)	Cost per unit	Cost for DM (c/kg)	Cost for energy (c/MJ)
Liquid feeds						
Whole milk						
–cheap	13	22.3	23	20 c/L	154	6.9
–expensive	13	22.3	23	30 c/L	230	10.3
Milk replacer						
–cheap	96	20.2	26	$60/bag	311	15.4
–expensive	96	20.2	26	$70/bag	364	17.9
Concentrates						
Pellets	90	13.0	18	$350/t	38	2.9
Farm mix	90	13.0	18	$250/t	27	2.1
Roughages						
Lucerne/clover hay	85	9.0	18	$3.5/bale	16	1.8
Cereal straw	90	7.0	3	$1.5/bale	7	1.0
Grazed pasture	20	11.0	14	$60/t DM	6	0.5

Other costs to consider in calf rearing

Surveys conducted in the US and the UK found that feed accounted for 50–60% of the total costs for raising heifer replacements to first calving. This proportion would probably be higher in Australia at the present time with its lower housing costs. However, Australian farmers would use more low-cost grazed pasture than their counterparts in North America and Europe.

In a Victorian study assessing the economics of pink veal production (Moran 1990), feed contributed to 53% of total production costs and calf purchase 27%, while the remainder was accounted for by management and labour (9%), transport (8%), power and repairs to rearing facilities (2%) and animal health (1%). These calves were fully housed and fed high concentrate diets until slaughter at 4–5 months of age, so the relative costs of feed and other components of production would differ from those in calves in less sophisticated rearing systems. The study did not account for capital costs for building the rearing shed or for any interest on borrowed money. Further details of the production costs and profit margins from different pink veal systems have been presented by Moran (1990) and will not be discussed in this chapter.

As previously discussed in Chapter 6, labour requirements have a large influence on decisions as to the most appropriate calf rearing system on dairy farms. In seasonal calving areas, farmers aim for minimum spread of calving and, hence, maximum concentration of calves to rear. This often coincides with other farm operations, such as haymaking, mating and early lactation feeding, and milk returns are generally lowest when milk supplies are highest. Therefore, daily time management would probably be given a high priority when planning rearing systems. In contrast, year-round calving herds provide a continual spread of calves to rear in much smaller numbers at any one time. Rearing facilities can be smaller and more sophisticated, and more time can be devoted to feeding heifer replacements.

It is not easy to quantify the benefits of the capital costs in providing adequate shelter for young calves. To fully house 85 calves all year round at Kyabram, a 20 m by 13 m shed consisting of 1.5 m high brick walls and blinds (for protection against wind and rain) with a reinforced cement floor was built in 1988 for $28,800. Simple calf hutches can be built for $60–$80 each, while cheap shedding would cost even less per calf. Shade trees and shelterbelts of hedges may provide adequate protection for most of the year, but without better protection during wet windy periods, animal performance will suffer, mortality rates could dramatically increase and profitability fall. The comparison of rearing systems in Table 14.3 does not include the actual cost for shelter, although it does take into account depreciation of facilities. Capital costs and/or their depreciation should always be considered when preparing budgets for rearing systems.

Budgets for calf rearing should also include some estimate for operating the rearing unit. Such operating costs would include power for lights and heating water and water for drinking and cleaning, as well as replacement costs of teats, buckets and other feeding equipment. Losses due to calf mortalities and costs for veterinary treatment and drugs should also be included, as well as an assumed mortality rate of 3 or 4% of reared calves. Some economists include the opportunity cost of the calf, or at least the interest of the total value of calves being reared at any one time. If animals have to be bought as week-old calves while others may be sold as weaned calves, transport and selling charges should also be included in any budget.

Categorising calf and heifer rearing costs in the US

A study of 62 dairy farms in Wisconsin has produced an interesting set of data on costs involved in calf and heifer rearing. The following data are all in US dollars so costs should be nearly doubled to be equivalent to Australia. The researchers separated costs into four categories as follows:

- Feed: milk or milk replacer, concentrates, roughages.
- Variable: bedding, veterinary and drugs, mating, fuel and electricity, death losses and interest.
- Labour and management: $7/h for labour and $12/h for management.
- Fixed: return on equipment and facility investment, but not the initial value of the calf.

The calf rearing period was the most expensive period, at $2.69/day, of which 42% was for labour and management, 36% for feed, 14% for variable and 8% for fixed costs. This equated to a total cost of US$260/calf, ranging from $185–$435/calf. Calf rearing costs were higher for smaller herds ($3.41/day for <75 cows) than for medium ($2.39/day for 75–150 cows) or larger herd ($2.57 for >150 cows), mainly due to higher labour and management costs. However, there was a fairly even spread of labour and management costs between the three herd size categories, indicating that there can be labour-efficient calf rearing operations on farms of any size.

Following weaning, total rearing costs declined to $1.22/day until 100 kg live weight, then steadily increased with live weight to reach $2.07/day by 550 kg. However, at around mating, they increased slightly in breeding cost, 20c/day, and additional feed costing, 10c/day. The average cost from birth to calving was $1.61, but this varied from $1.24–$1.88 between the farms. This equated to US$1360/heifer, ranging from $922–$1807/heifer. Feed constituted 59% of the total daily costs, with the other three categories comprising 12–16%.

Assuming one labour unit worked for eight hours each day, the efficiency of using labour and management was calculated in terms of total hours required to rear one heifer from birth to calving and as heifers reared per hour or per labour unit per year. Benchmarks were then developed for the three herd-size categories. Unfortunately, benchmarks of the calf rearing operation have not been developed. The heifer rearing benchmarks were 9 hr/heifer (ranging from 7–12 for different herd sizes), 54 heifers/hr (ranging from 39–62) and 430 heifers/labour unit/year (ranging from 310–495). However, it is one thing to have a labour efficient operation, but is must also produce healthy productive heifers.

The cost of diseases in calves

The losses through disease during calf rearing can be separated as follows:

- Deaths, hence, loss of calf value with little (or usually no) salvage value for the carcass.
- Costs of veterinary services plus drugs.
- Costs of extra feed required when calves lose or do not gain weight when sick, hence, require more feed to reach target live weights.
- Costs of transport and resale of any calves culled.
- Costs of reduced throughput in rearing unit, additional labour for treatment and greater interest on loans, etc.

The relative significance of these various costs of disease were studied in calf rearing units in England and reported by Webster (1984). For a 100-calf-rearing unit in which calves cost the equivalent of $160 to buy and were sold for $300 at 12 weeks of age, the gross profit margin with no deaths or disease was $6200. With 'normal' deaths and diseases occurring (5% mortality and 13% calves recovering after veterinary treatment), this margin was reduced to $4400. However, when an outbreak of pneumonia killed 21 calves of the 100 calves and another 14 calves recovered after treatment, the margin was reduced to only $440. This clearly shows that the veterinary and extra food costs of keeping calves alive to 12 weeks of age are trivial to the losses incurred when calves die.

Taking into account the costs of treatments, loss of growth, depressed sales of poorly growing calves, delays to first services and labour costs, an outbreak of pneumonia can cost UK£38/calf (Esslemont and others 1998), while an outbreak of scours can cost £33/calf (Gunn and Stott 1998).

Veterinarians in the US have calculated that each sick calf requires, on average, 53 minutes of extra care before recovery occurs. In terms of labour, veterinary services and drugs, the cost for each sick calf is at least US$18. Good calf rearing and husbandry and sound economics must then go hand in hand.

A case study of cost savings through changing milk feeding systems

In recent years there has been an increasing awareness of savings that can be made in seasonal calving herds through storing all transition milk for use later in the calf rearing season. Table 14.2 presents such a case study from Gippsland in south-eastern Victoria, where a dairy farming couple milk 370 cows.

In 1994, they reared 85 calves in several small paddocks on whole milk, feeding each calf 6–8 L/day, plus ad lib concentrates, until 8 weeks of age. Any calves weighing less than 75 kg were then fed 4 L/day for three more weeks. In 1995, they reared 105 calves in a converted hay shed, with up to 15 calves per pen, on 4 L/calf/day of whole milk plus ad lib concentrates and straw. Calves were weaned off milk when eating 1 kg/day of concentrates, which occurred by 5 weeks of age but kept in the shed, then given access to pasture two weeks later. The total feed requirement per heifer calf to 12 weeks of age in 1994 was 490 L milk (which included 375 L saleable milk) plus 12 kg concentrates, while in 1995, each calf consumed 140 L milk (which was all non-saleable milk), 58 kg concentrate plus 20 kg straw.

Table 14.2 Case study of Gippsland dairy farm where the producers changed their calf rearing system between 1994 and 1995

Year	1994	1995
Heifer calves reared	85	105
Weaning age (wk)	8–11	5
Average milk intake (L/calf)	490	140
Market milk fed (L/calf)	375	–
Concentrate intake (kg/calf)	12	58
Straw intake (kg/calf)	–	20
Total feed costs ($/calf)	79.2	23.3
Total calf rearing costs ($/yr)	6732	2656

The change in calf rearing system resulted in calves in 1995 being milk reared entirely on colostrum and transition milk. In Table 14.2, market milk was valued at 20 c/L, concentrates at $350/t and straw at $3/20 kg bale. In 1994, total feed costs were $79.2/calf or $6732 for 85 calves, compared with only $23.3/calf or $2446 for 105 calves in 1995, a saving of $4286. The 1995 calves were also quieter to handle, healthier and had a more even range of live weights at 12 weeks of age. The farmer considered that by 4 months of age, there was little difference in live weight between the two batches of calves.

This example clearly indicates the savings that can be made through utilising all available milk for rearing replacement heifer calves.

Comparing different systems to calculate total feed costs for the first 12 weeks of rearing

The development of a computer spreadsheet (KYHEIF, Moran unpublished data) has allowed the simulation of different calf rearing systems with varying inputs of transition milk, whole milk or milk replacer, concentrates and roughages, up to 12 weeks of age. The spreadsheet was developed using mathematical relationships to predict the energy requirements of milk-fed and early weaned calves for maintenance, activity and growth, published by Roy (1980), MAFF (1984) and AFRC (1993).

The nutritive value (fat and protein content of liquid feeds, energy and protein contents of solid feeds) and costs of the various feeds used during the first 12 weeks of

Economics of calf rearing | 181

Figure 14.1 Individual rearing in the calf paddock – filling the buckets

Figure 14.2 Individual rearing in the calf paddock – feeding the calves

rearing are imputed in the spreadsheet. These are whole milk, milk replacer, concentrate (which is assumed to be the same pre- and post-weaning), roughage (only fed pre-weaning) and grazed pasture (only offered post-weaning). The weaning live weight is calculated from the birth weight, the age at weaning and the pre-weaning growth rate. The post-weaning growth rate is calculated using these data and the 12-week live weight. The concentrate intake pre-weaning is calculated from the daily milk intake and pre-weaning growth rate, with the roughage intake set to 200 g/day. The post-weaning intakes of concentrate and grazed pasture are calculated from their relative proportions of total DM intake. The feed costs are calculated from energy contributions of each feed type together with their energy costs, during both the pre- and post-weaning phases of growth. Total costs for liquid feeds are adjusted using the proportion of non-saleable transition milk, which can be imputed into the spreadsheet.

Table 14.3 presents a series of simulations for three different milk rearing regimes using various proportions of transition milk, based on the following assumptions:

- Calves weighed 35 kg at birth and 100 kg at 12 weeks of age.
- Whole milk (3% protein and 4% fat) was valued at 30c/L.
- Concentrates (12 MJ/kg DM of energy and 18% crude protein) cost $350/t.
- Roughage (straw) cost $60/t.
- Grazed pasture (10.5 MJ/kg DM of energy) cost $60/t DM.
- The post-weaning diet constituted 70% concentrate and 30% grazed pasture.

Table 14.3 Feed requirements and costs to rear calves to 100 kg at 12 weeks of age, when weaned at 6, 8 or 10 weeks, assuming 50% of the transition milk is available for feeding

Weaning age (weeks)	6.0	8.0	10.0
Details of regime			
Milk intake (L/day)	4.0	6.0	7.0
Pre-wean growth rate (kg/day)	0.4	0.75	0.8
Weaning live weight (kg)	52.0	77.0	94.0
Feed inputs per calf			
Total milk (L)	168.0	336.0	490.0
Vat milk (L)	98.0	266.0	420.0
Total concentrate (kg)	120.0	65.0	23.0
Straw (kg DM)	8.0	11.0	14.0
Grazed pasture (kg DM)	52.0	28.0	10.0
Total feed costs ($/calf)			
Pre-weaning ($)	25.2	68.2	105.7
Post-weaning ($)	44.2	23.8	8.8
Total ($)	69.4	92.0	114.5
Total feed costs at different % transition milk ($/calf)			
30% ($)	77.0	103.0	128.5
70% ($)	61.8	81.0	100.1

Total feed costs for feeding milk until 10 weeks were $45/calf (or 65%) above those if weaning calves at 6 weeks of age. Increasing the proportion of transitional milk available for heifer replacements reduced total feed costs by up to $28/calf in any one weaning system. The difference between the cheapest and most expensive milk rearing regimes in Table 14.3 is more than $67/calf (or 108%).

References and further reading

Agricultural Food Research Council (1993), *Energy and Protein Requirements of Ruminants*, CAB International, Wallingford, UK.

Esslemont, R., Kossibati, M. and Reeve-Johnson, L. (1998), 'The Costs of Respiratory Diseases in Dairy Calves', p.685–90, *Proc XX Wld. Assoc. Buiatrics Cong.*, Sydney.

Gunn, G. and Stott, A. (1998), 'A Comparison of Economic Losses Due to Calf Enteritis and Calf Pneumonia in Scottish Herds', p.357–60, *Proc XX Wld. Assoc. Buiatrics Cong.*, Sydney.

Ministry of Agriculture, Fisheries and Food (1987), *Feed Composition. UK Tables of Feed Composition and Nutritive Value for Ruminants*, Chalcombe Publications, Marlow, England.

Moran, J. (1990), *Growing Calves for Pink Veal. A Guide to Rearing, Feeding and Managing Calves for Pink Veal Production in Victoria*, Vic. Dept. Agric. Tech. Rep. 176, Melbourne.

Roy, J. (1980), *The Calf*, Fourth Edition, Butterworths, Sydney.

Webster, J. (1984), *Calf Husbandry, Health and Welfare*, Granada, Sydney.

fifteen 15

Best management practices for rearing dairy replacement heifers

What makes a good calf rearing system?

For every hundred dairy farmers, there would be close to a hundred different ways they rear their heifer replacement calves. More and more farmers are feeding less milk for fewer weeks and reaping the benefits of early rumen development, provided they feed top quality concentrates and low quality, but palatable, roughages. Most farmers now provide shedding for their milk-fed calves to protect them from the extremes of climate. There may be only 20 different rearing 'systems', but a lot more rearing practices, after taking into account the subtleties of physical facilities, feeding programs, disease management and human/calf interactions.

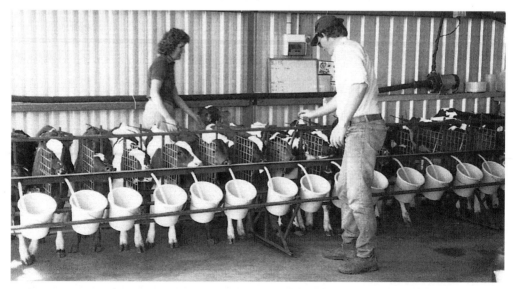

Figure 15.1 A well-managed, cost-efficient calf rearing system

A survey of 1700 dairy farms in the US documented a series of calf rearing practices that were closely associated with mortality in milk-fed calves. For a variety of reasons, US dairy farmers lose many more pre-weaned calves than do Australian farmers. Mean death rates in the US are 8–9%, compared to only 3–5% in Australia. The following list summarises those factors that were associated with high death rates in US calf rearing units in year-round calving herds. We can then assume that the converse of each detrimental factor indicates a beneficial effect on calf health and survivability.

It must be emphasised that these data are for 'typical' US dairy systems that differ in many ways to Australian farms. Therefore, it is the general principles rather than the actual 'numbers' that are important to us in this study.

Herd performance

Low producing herds have higher calf losses. In the study, low production was quantified as less than 7700 L/cow rolling herd average, which would equate to something like 3500–4000 L/cow in Australia. One concludes that lower producing herds were less carefully managed, including the calf-rearing unit.

Size of operation

Larger calf rearing units had higher mortality rates. A large unit is defined as one rearing more than 30 calves over a three-month period, which would equate to 120 calf/year unit on a year-round calving herd. One concludes that bigger units are less well managed than smaller units.

First colostrum feeding management

Farmers allowing their newborn calves to suckle their dam for their first drink of colostrum have higher death rates than do those feeding it by hand. Removing the calf from her dam immediately after birth reduces the chances of ingestion of faecal material as the calf looks for the teat. Furthermore, a controlled feeding of sufficient, good quality colostrum within a short period after birth will ensure absorption of sufficient colostral antibodies into the calf's blood stream. Depending on nature to do this in a modern day milking herd is a lot more haphazard.

Group size

Calves reared in groups of seven or more have higher death rates than calves reared in groups of six or less. The smaller the group size, the better individual attention for each calf.

Gender of rearer

Calves reared by men had higher death rates than calves reared by women. One can only conclude that women have better rearing skills than men. In my experience, the best calf rearers are women nurses, since they have been trained to anticipate health problems before they happen.

Relationship of rearer to farm owner

Calves reared by the farmer's children or employees have higher death rates than calves reared by the farm owner or spouse. One can conclude that if you own the calves, you will take more care with them.

Time of feeding roughage

Higher death rates were found on farms where feeding of hay or other roughages was delayed until calves were 20 days or more old. The earlier that calves are offered roughage, the sooner their rumen begins to develop and the sooner they are likely to be weaned, either voluntarily or because of the feeding system. There was no effect of age of feeding concentrates or free choice of water on calf death rates.

Feeding mastitic or antibiotic milk

Heifer calves reared in units where mastitic or antibiotic milk was fed to them had higher death rates then in units where it is discarded, or fed to less valuable calves.

Feeding whole milk to calves

Calves fed whole milk from a bulk tank had lower death rates than calves not fed whole milk. It is assumed that calves not fed whole milk were fed milk replacer.

These nine factors sum up a good calf rearing system, in which calves are provided with close attention to their health, digestive development and welfare by a person who really cares for them.

Figure 15.2 Dairy advisers should be the first point of call when planning new rearing systems

Monitoring your calf rearing system

It has been frequently stated, 'If you cannot measure it, you cannot manage it'. In the course of their operations, calf rearers already do collect, and can easily collect more, data on their stock. Much of this data could be, but probably has not been, used in making management decisions to improve the profitability and efficiency of their operation. Such decisions include:

- By recording the ear tag of any calf requiring veterinary treatment then recording how many lactations it remains in the milking herd, decisions could be made on whether treated calves should be sold or still kept as replacements in the herd.
- Once farmers know the total costs for their first-calf heifers to enter their dairy herd and start generating income, they can then decide on whether it is more profitable to sell all their calves and rely on purchased in-calf heifers to maintain or expand herd sizes.
- By monitoring feeding and management costs from weaning to first calving, farmers can compare that with agistment or contract heifer rearing.
- By monitoring live weights and wastage rates, at different stages of rearing, farmers can decide on optimum target live weights and, hence, feeding management for their particular operation.
- As milk factories require farms to enter audits for quality assurance programs, some record keeping will be mandatory.
- Record keeping can assist with identifying areas requiring attention and to direct staff to problem areas or potential risk areas on farms.
- Key measures, such as wastage rate (from birth to second calving) and heifer replacement rate (percentage of heifer calving per 100 cows bred), are major determinants of herd profitability.

The following lists some of these measures that can be easily collected and used in making these future decisions:

Pre-calving (heifer's dams)
- Genetic merit of dams of replacement heifers (Australian Breeding Value, or if herd testing, Production Index).
- Costs of semen and, hence, each live heifer calf.

Post-calving (heifer's dams)
- Percentage of calving difficulties.
- Percentage of calves born dead.
- Colostrum quality (percentage of different quality categories).
- Calf antibody status (percentage above 8 mg Ig/mL).

Pre-weaning

- Litres of vat milk used to feed each calf.
- Weekly concentrate intake; as a guide to weaning age.
- Percentage of calves that die, are sick or sold (and why).
- Record ear tag of each treated calf; to assist in future decision making about their fate.
- Average weaning age.
- Approximate time spent on rearing calves (hours/day and hence minutes/calf until weaning).
- Costs of purchased feeds.
- Costs of veterinary treatment (drugs and visits).
- Costs for routine management (vaccines, drenches, etc).
- Capital cost of shed and equipment; to calculate costs for depreciation.
- Live weight and wither height at 12 weeks of age; to compare pre-weaning performance from year to year.

Pre-mating

- Weekly concentrate and hay inputs; to help plan future feeding programs as they vary with season.
- Quality of supplements; this should include vendor declarations of purchased feeds as well as measures of their nutritive value.
- Live weight and wither heights at 6, 9, and 12 months and at mating at 15 months.
- Conception rate at mating.
- Inseminations per conception, if using artificial insemination.
- Faecal egg counts at strategic times, to assist with drenching program.
- Percentage of heifers that die, are sick or sold (and why).
- Costs of purchased feeds, veterinary treatment, routine management.
- Costs of mating (semen and oestrus synchronisation or bull).
- Total rearing costs per calving heifer; the 'bottom line'.

Post-calving

- Days to successful insemination.
- Inseminations per conception.
- Percentage of first lactation heifers that die, are sick or sold (and why).
- First lactation yield of milk or milk solids.
- First lactation yield as percentage of yields on mature cows.
- Wastage rate from birth to second calving, as percentage of heifer calves reared.
- How many lactations remaining in the milking herd.
- When eventually culled, for what reason.

What is best management practice and quality assurance?

Best management practice (BMP) and quality assurance (QA) are processes for describing and implementing the most suitable procedures for a particular set of tasks to achieve a desirable outcome. With something as complex as operating a dairy enterprise, it is best to partition the major outcome, that is profitable milk production, into several sets of management decisions that producers must make, such as growing productive pastures, grazing management, effective animal health and rearing replacement heifers. Essentially, BMP means:

- Saying what you do.
- Doing what you say.
- Recording what you have done.
- Hence, improving that aspect of your dairy operation.

Quality assurance (QA) programs are in their infancy in Australia, although some are currently being implemented by other non-dairy livestock producers, such as Clipcare and Flockcare in sheep, Cattlecare in beef and forage vendor declarations by the Australian Fodder Industry Association. With increasing focus on meeting customer requirements, QA programs will become an integral part of dairy enterprises in future years.

Although heifer rearing programs must be tailored towards individual producers, there are several general principles associated with all BMPs. Those relevant to heifer rearing programs, which producers should aim for, are:

- Incorporating heifer rearing into a business plan for the entire dairy enterprise.
- Making a commitment to continuous improvement in rearing costs, timeliness of each phase of the program and on the end product, namely heifer quality.
- Developing closer relationships and alliances with all outside service providers to the program, such as veterinarians, feed suppliers, dairy advisers, semen suppliers and AI technicians.
- Using performance recording, then benchmarking your achievements with the industries' best, for performance indicators such as heifer wastage rates, heifer milk production (as a proportion of herd average), heifer fertility and cost per first lactation heifer.
- Integrating environmental and animal welfare concerns in all aspects of the program.
- Being involved with other producers to improve your knowledge and upgrade the competitiveness of your dairy enterprise.

The Milk and Dairy Beef Quality Assurance Program is a national program developed in US (Dairy Quality Assurance Centre 1998). Its major objective is to increase herd performance while at the same time reducing herd costs and meeting customer demands for animal care and quality animal products. One of the modules of the total cattle management plan is entitled 'Quality replacement heifers – Growing your profits'. This consists of various checklists to assist producers in developing BMPs for their heifer replacement program. The checklists document the most appropriate questions dairy farmers, their staff, consultants and suppliers of services to the heifer rearing program should seek to answer. The audits are designed to encourage implementing a team approach to quality management.

This audit was developed on accepted norms for US dairy producers. However, several of these practices have yet to be accepted as routine by Australian producers, such as testing all cows for Johne's disease or enzootic bovine leucosis. Nevertheless, such practices have been included in the following checklists, since in the future, they may be integrated into Australian dairy production systems.

Several of the checklists contain procedures that are outside the scope of this book, such as mating management of dairy heifers. However, as they are an integral part of any heifer operation, they have been included in this chapter.

The following checklists can form the basis of a QA scheme for rearing dairy replacement heifers in Australia. Each one can be enlarged and placed on a notice board in the dairy office to remind producers that continually improving their young stock management as one key area of their overall farm management.

Checklists for quality assurance when rearing dairy replacement heifers

1. Planning general herd and heifer management

Profit from quality starts and ends with a commitment by owners and/or managers of dairy enterprises to seek this quality. Unless management is committed to improve quality, few gains will be achieved by producers and service providers associated with heifer operations. A commitment to producing quality replacement heifers is one important step towards increasing herd health and quality milk production.

Producers should review the following checklist in Table 15.1 to see how many of the 'Yes' boxes they can tick

Table 15.1 A checklist for general herd and heifer management

Yes	No	Best management practices
☐	☐	Do you consistently meet milk quality standards for somatic cell counts and bacterial levels?
☐	☐	Do you believe it is more profitable to increase milk quality and yield via replacement heifers than to improve milk quality and yield from the current herd?
☐	☐	Do you have a permanent 'tamper proof' animal identification system in place?
☐	☐	Can you readily track and validate to others the quality represented in your replacement heifers?
☐	☐	If you rear your heifers off farm, do you have in place a measurement system to evaluate such a rearing program?
☐	☐	Have you been able to consistently produce more milk per cow each year?
☐	☐	Have you been able to consistently increase milk production per hectare of grazed pasture each year, after allowing for purchased supplements?
☐	☐	Do your first lactation heifers consistently produce >80% of the full lactation milk yield of your mature cows?
☐	☐	Is your target live weight at first calving based on breed and target milk yield of your mature cows in the herd?
☐	☐	Do you participate in other QA audits in a quality assurance program?

2. Planning heifer supply programs

Quality replacement heifers programs can be described as those that produce strong healthy heifers at 24 months of age, after which they become productive members of the milking herd for at least five lactations. Planning quality heifers starts with the present herd. The key time to start 'quality' replacement heifer programs is before mating.

Producers should review the following checklist in Table 15.2 to see how many of the 'Yes' boxes they can tick

Table 15.2 A checklist for planning heifer supply programs

Yes	No	Best management practices
☐	☐	Do you participate in a herd testing program to help identify the best cows from which to produce your replacement heifer calves?
☐	☐	Have you and your veterinarian developed a routine herd health program including appropriate vaccination schedules for your cows?
☐	☐	Do you routinely score the body condition of your cows to evaluate management and assure that they are in good condition to produce thrifty, healthy calves?
☐	☐	If seasonally calving, do you have a compact calving program with all your replacement heifer needs born within 6–8 weeks?
☐	☐	If not, are you able to achieve this by using AI (artificial insemination) over your maiden heifers?
☐	☐	Do you select sires that will breed top quality replacement heifers?

3. Planning heifer care from birth to weaning

Quality replacement heifers start by being strong calves at birth, followed by a quality meal of colostrum and then consuming solid feed by 5 weeks of age.

Producers should review the following checklist in Table 15.3 to see how many of the 'Yes' boxes they can tick.

Table 15.3 A checklist for planning heifer care from birth to weaning

Yes	No	Best management practices
☐	☐	Are your calf mortality losses at birth less than 5% of calves born?
☐	☐	Are your calf mortality losses from birth to weaning less than 3% of live calves born?
☐	☐	Are your springing cows and heifers provided with a clean dry area for calving?
☐	☐	Do you remove the calf from her mother, preferably at birth, but at least within the first 12 hours?
☐	☐	Do you dip the navel in a strong (7%) iodine solution immediately after birth?
☐	☐	Do you have a good program to supply high quality colostrum, such as: 1. Use a colostrometer to monitor colostrum quality and only feed good quality colostrum?
☐	☐	2. Ensure the calf gets two or more litres immediately after birth?
☐	☐	3. Provide another two or more litres within 6–12 hours of birth?
☐	☐	4. Use a stomach tube, if necessary?
☐	☐	5. Clean bottles, buckets and equipment regularly?
☐	☐	6. Pool colostrum from older cows, tested negative for Johne's disease and enzootic bovine leucosis?
☐	☐	Do you blood test some calves to check the efficiency of your colostrum program?
☐	☐	Do you remove calves to a clean dry area, preferably at birth, but at least within the first 12 hours?
☐	☐	Do you use a permanent form of identification for each calf?
☐	☐	Do you minimise contact between batches of calves until about 5 weeks of age?
☐	☐	Do you provide calves with access to water at all times?
☐	☐	Do you provide at least 1.5–2 m^2/calf during milk rearing?
☐	☐	Do you provide concentrates to each calf within the first week of age?

☐	☐	Do you ensure calves each consume 0.75 kg of calf concentrate for two or more days prior to weaning
☐	☐	Do your calves weigh 100 kg (Friesian) or 90 kg (Jersey) by 12 weeks of age?
☐	☐	Do you weigh calves using cattle scales?
☐	☐	Do you monitor the health and welfare of calves at least twice each day?
☐	☐	Have you discussed your calf health management program with your veterinarian?
☐	☐	To minimise the spread of Johne's disease, do you remove all possible avenues of infection between adult animals and calves?
☐	☐	Do you isolate any calves showing signs of ill health and to minimise spread of infection, then feed them last?
☐	☐	Do you know how much money early weaning can save?
☐	☐	Do you quarantine any pre-weaned calves introduced onto your farm?
☐	☐	Do you use the best possible feeds in your program (milk replacer, concentrates, straw)?
☐	☐	Have you discussed the use of waste (antibiotic/mastitic) milk with your veterinarian?
☐	☐	Do you dehorn and remove extra teats from heifer calves during the milk rearing stage?
☐	☐	When selling excess calves, do you follow all the regulations regarding suitability for sale (minimum age, health status, antibacterial residues) and transport them in a suitable trailer?
		Score each of the following six disease problems you may encounter during milk rearing.
		Use a ranking of 1 (least problem) to 6 (biggest problem)
____		Scours or diarrhoea
____		Respiratory problems
____		Joint or navel problem
____		Trauma
____		Unknown
____		Rarely have illness

4. Planning heifer care from weaning to mating

Quality heifers are a challenge between weaning and mating. Once they have been weaned off milk, their feeding management is frequently neglected, yet this is one of the most important periods of their life. Heifers achieve puberty at about one half their mature size. The sooner they reach puberty and start cycling, the more likely they will conceive when mated at 15 months of age.

Producers should review the following checklist in Table 15.4 to see how many of the 'Yes' boxes they can tick.

Table 15.4 A checklist for planning heifer care from weaning to mating

Yes	No	Best management practices
☐	☐	Do you have feeding strategies to minimise the growth check immediately after weaning?
☐	☐	Does your focus on heifer growth include nutrition, health, parasite control and social factors?
☐	☐	Do you understand the importance of achieving target minimum live weights and wither heights for age at every stage of heifer development?
☐	☐	Do you routinely monitor heifer growth using cattle scales?
☐	☐	Do you routinely monitor wither height?
☐	☐	Do you understand the change heifers go through as they switch from a milk-based diet to a fully developed ruminant utilising solid feeds?
☐	☐	Do you feed calf concentrates before, during and after weaning?
☐	☐	Do you base your feeding program on growth rates, which can vary dramatically with the availability and quality of pastures being grazed?
☐	☐	In year round calving herds, do you group heifers on age and/or live weight?
☐	☐	If growth rates fall below acceptable targets, do you supplement heifers with quality feeds, such as cereal grain and/or good quality conserved forages?
☐	☐	Are you aware of potential problems (fatty udder syndrome) arising from feeding excess high energy/low protein feeds between weaning and puberty?

☐	☐	To minimise the spread of Johne's disease, do you ensure heifers under 12 months of age do not graze pastures that have been stocked with adult animals during the previous 12 months?
☐	☐	Have you developed a health management program (vaccinations, internal and external parasite control) in conjunction with your veterinarian?
☐	☐	Do you use individual needles during any vaccination program requiring intramuscular injections, disinfecting needles in alcohol?
☐	☐	If your herd is diagnosed with enzootic bovine leucosis, are you meticulous in ensuring no cross contamination of animals with blood or milk?
☐	☐	Do you have an effective fly control program, if necessary?

5. Planning heifer mating programs

Successful mating programs for replacement heifers require all animals to be cycling. All the hard work and quality management will only return profit to your operation if conception rates are high when heifers are mated at 15 months of age.

Producers should review the following checklist in Table 15.5 to see how many of the 'Yes' boxes they can tick.

Table 15.5 A checklist for planning heifer mating programs

Yes	No	Best management practices
☐	☐	Do your replacement heifers have body condition scores of 5–6 units and are they gaining weight at mating?
☐	☐	By 15 months of age, have all your replacement heifers achieved minimum target live weights for mating (330 kg for Friesians and 245 kg for Jerseys)?
☐	☐	Do you treat your heifers for internal and, if required, external parasites just prior to mating?
☐	☐	With seasonal calving herds, do you plan heifer calving dates in relation to those of your mature cows? This may be a week or two earlier to assist with feeding management of newly calved heifers.
☐	☐	Are you aware of the benefits of using AI (artificial insemination) over natural mating?
☐	☐	Are you aware of the benefits of using dairy (as against beef) bulls or semen?
☐	☐	Do you use AI for mating well-grown heifers, then follow on with good quality herd bulls to 'clean up' these heifers and any smaller ones not artificially inseminated?
☐	☐	If using AI, do you select semen from sires or breeds selected for ease of calving?
☐	☐	Have you selected the most appropriate heat detection procedure for your operation?
☐	☐	Do you choose to use heat synchronisation, if appropriate, for your heifer mating program?
☐	☐	If using natural mating, do you select sufficient good quality bulls, taking note of their mobility and libido (one bull per 30 heifers plus one spare)?
☐	☐	If using natural mating, do you ensure all bulls have been vaccinated against vibriosis?
☐	☐	Do you routinely pregnancy test your heifers to plan their calving program?

6. Planning heifer care from mating to calving

Good heifer management is vital up to the point of calving, particularly if the calf is destined for the replacement herd.

Producers should review the following checklist in Table 15.6 to see how many of the 'Yes' boxes they can tick.

Table 15.6 A checklist for planning heifer care from mating to calving

Yes	No	Best management practices
☐	☐	Do your heifers gain on average 0.6–0.8 kg/day after mating?
☐	☐	Do your heifers calve down in body condition score 5–6 units?
☐	☐	Do you introduce your heifers to the milking shed (or at least run them through the milking shed) prior to calving?
☐	☐	Do you avoid mixing replacement heifers with older dry cows?
☐	☐	Do you avoid high somatic cell counts by keeping replacement heifers in a clean dry paddock for at least one month before calving?
☐	☐	Do you store good quality, tested colostrum from older cows to routinely feed all calves from first-calf heifers (either freshly chilled or frozen from the previous year)?
☐	☐	Have you and your veterinarian developed a health treatment program for heifers pre and immediately post calving?

References and further reading

Dairy Quality Assurance Centre (1998), *Quality Replacement Heifers. Growing Your Own Profits*, Iowa, US.

Losinger, W. and Heinrichs, A. (1997), 'Management Practices Associated with High Mortality Among Preweaned Dairy Heifers', *J. Dairy Res.*, 64, 1–11.

John Moran's 10 golden rules of calf rearing

1. Ensure each calf receives 4 L top quality colostrum within six hours of birth. Remember the three Qs for colostrum feeding (quality, quantity, quickly). Dip or spray the umbilical cord with iodine solution.
2. Remember that feeding milk only once each day encourages faster rumen development and reduces rearing costs, ensuring less health problems and better post-weaning performance.
3. Provide continual access to clean water and high quality concentrates from day one. Also provide a palatable roughage source, such as clean straw.
4. Give individual attention to each calf and make time to check at least twice daily for signs of ill thrift or sickness.
5. Develop a disease action plan that includes good hygiene, isolation of sick calves, fluid replacement and TLC. Drugs should only be used a last resort, to complement a well-managed system.
6. Milk rear calves in a clean, dry, well-ventilated shed, in groups no more than 10 animals, providing at least 1.5 m^2/calf.
7. Ensure good record keeping, as this will help pin-pointing problems in your system.
8. Minimise stresses through following set routines each day, reducing over crowding and 'keeping your troubles' out of the calf shed.
9. People rear calves, not systems! Do not rear calves if you don't enjoy it – find a specialist.
10. A good calf rearing system produces healthy, fully weaned calves weighing 100 kg at 12 weeks of age.

appendix 2

John Moran's golden rules of heifer rearing

The principles of good heifer rearing can be summarised in the following key points:

Targets

- Ranges of target live weights and wither heights for ages with Friesian heifers in well-managed herds are:

Age (months)	Live weight (kg)	Wither height (cm)
3	90–110	88
12	270–300	118
15	330–360	122
24	520–550	135

- The optimum pre-calving live weight of heifers varies with their target milk yield as mature cows. In Friesians, this can range from 500 kg in herds averaging 5000 L/cow/year to 560 kg in herds averaging 7000 L/cow/year.
- During their first lactation, well-reared heifers should produce at least 80% of the full lactation milk yield of their mature herd mates.
- Heavier heifers must be fed well to achieve their economic benefits. There is little point in growing out heavy heifers then underfeeding them as milkers.
- Heifers should be managed to grow at an average of 0.7 kg/day from weaning to first calving, although this can vary during the 24 months from 0.5–1.0 kg/day depending on seasonal conditions.

Feeding

- Heifers should be provided with a good quality diet for their first 12 months, containing 10–11 MJ/kg DM of energy and 12–16% protein.
- High energy supplements are often required to achieve target growth rates of young heifers (up to 6 months of age), particularly during their first winter.
- Grazed pasture is generally the cheapest feed, but it must be in sufficient supply – at least 1800 kg DM/ha.
- If insufficient pasture is available, heifers should only be fed top quality supplements, preferably concentrates.
- Any hay or silage fed must be of good quality, at least 10 MJ/kg DM of energy and 14% protein.
- Be wary of feeding an unbalanced diet containing too much low protein grain during the 3–9 months of age critical period, as excessive growth rates can lead to fatty udders and reduced milk potential.

Management

- If young stock are allowed to lose weight or grow very slowly for lengthy periods, they will not achieve their potential frame size.
- Low mating live weights can lead to calving difficulties nine months later. Excessive feeding after mating can also result in dystocia. Dystocia reduces milk yield and increases the number of days to the second conception.
- Use AI and quality Friesian semen on well-grown heifers to provide replacement calves from first-calf heifers.
- Heifers should be regularly weighed, at least every three months, with wither heights recorded at each weighing.
- Contract heifer rearing costs $4.50–$6.50 per week. When considering this option, producers should take into account all the costs of rearing their heifers on-farm, such as feed, labour, health and using pasture otherwise available for milking cows.

appendix 3

Glossary of technical terms

Abomasum
The fourth (or true) stomach in ruminant animals that can digest feeds using enzymes produced in the stomach wall.

Acid and alkaline
Measures of the pH in gut contents. Different parts of the digestive tract require acid or alkaline conditions for optimum digestion of feed.

Acidosis
This grain poisoning occurs when rumen pH falls too low through over production of lactic acid, reducing feed digestion and sometimes causing death.

Ad lib
A term meaning feeding calves to full appetite.

Amino acids
The 'building blocks' of protein to which feed protein is broken down in the digestive tract.

Anaemia
A condition caused by very low levels of iron in blood and body tissues.

Antibiotics
Drugs, generally prescribed by veterinarians, to treat diseases by killing specific bacteria. Unfortunately, their use is becoming too prevalent in normal calf rearing, such as their inclusion in commercial milk replacers.

Antibiotic residues
Antibiotics remaining in animal products when sold for human consumption.

Antibodies
Proteins produced by animals in response to specific diseases. These are passed on to newborn calves through the colostrum.

Antioxidants
Included in milk replacers to reduce deterioration of fat during storage.

Anti-trypsin antigen
A chemical in untreated soya flour, which upsets milk digestion in calves.

Best management practice (BMP)
A description of the most suitable procedures for undertaking a set of tasks.

Biological value
A measure of the value of feed protein for use by animals.

Biosecurity
The protocols introduced to minimise the introduction of diseases into the calf shed.

Bloat
A condition caused by over distension of the abomasum or rumen, which requires immediate attention as it can quickly kill animals.

Buffers
Chemicals that stop sudden changes in rumen pH.

Burdizzo
A bloodless method of castrating bull calves by crushing the cord of the testes.

Calf scales
Selling points for week-old calves where producers are paid a set amount depending on calve live weights. They can be fixed or mobile, in which case trucks visit farms to weigh and collect the calves.

Casein
The major protein in milk products. It is the only one that can be completely digested by very young calves.

Clostridia
Bacteria causing a variety of diseases in calves and older cattle.

Coccidia
Microbes called protozoa, which cause scouring in calves.

Colostrometer
A device that measures the level of immunoglobulins (Ig) in colostrum. It is sometimes called a colostradoser.

Colostrum
Colostrum or beastings is the milk produced by cows for the first two milkings post-calving, which contains high levels of nutrients and immunoglobulins (Ig) for transferring immunity onto newborn calves.

Crowding disease
A general term for influenza-type respiratory diseases.

Cryptosporidia
Microbes called protozoa, which causes scouring in calves.

Degradability
A measure of the degree of breakdown of dietary protein by rumen microbes.

Duodenum
The first section of the small intestine.

E. coli
Bacteria, causing scouring in calves.

Electrolytes
Mineral salts used to alter the pH of gut contents for optimum digestion. Electrolyte solution is a solution of salts (and often an energy source such as glucose) used to replace fluids lost during scouring.

Enterotoxaemia
Enterotoxaemia or pulpy kidney is one of the clostridial diseases.

Enzymes
Chemicals produced by animals that assist with the breakdown of feeds in the digestive tract; pepsin, lactase, rennin, lipase, galactase are all enzymes.

Fatty acids
The end products of digestion of fats in the diet.

Five-in-one
A vaccine used to protect against clostridia bacteria.

Flight zone
The personal space around animals where they will attempt to move away from people.

Gossypol
A toxin in cottonseed that can kill calves (but not cows).

Haemoglobin
The chemical in the blood containing iron, used to store oxygen for release into body tissues.

Immunoglobulins (Ig)
Blood proteins included in colostrum that pass on passive immunity to newborn calves.

Induced calves
Calves prematurely born to reduce the spread of the calving season in certain dairy regions. They are generally more susceptible to diseases.

Infectious bovine rhinotracheitis (IBR)
A respiratory disease in calves caused by a virus.

International unit (IU)
The measure of the concentration of vitamins in feeds.

Joint-ill
A bacterial infection of the umbilical cord in newborn calves, which can cause arthritis of the joints. It is also referred to as navel-ill.

Johne's disease
A bacterial infection of the intestines that is easily transmitted to young calves. It is a notifiable disease requiring strict control measures.

Lactic acid
An end product of cereal grain digestion that can cause acidosis.

Lactose
A major energy source from milk products.

Lecithin
Included in milk replacers to assist with the incorporation of added fats.

Leptospirosis
A bacterial disease that is prevalent among dairy farmers due to its transmission from cows and calves.

Medicine disease
Caused by prolonged use of antibiotics, which can upset the balance of rumen microbes.

Metabolisable energy
The feed energy available for animal growth after accounting for digestion and metabolism losses.

Microbial protein
An important component of the microbes in the rumen, which is broken down into amino acids for use by the animal.

Neonatal diarrhoea
A technical term for calf scours.

Nurse cows
Cows used for multiple suckling calves, either by running with them in at pasture (continuous or foster suckling) or by held-in specially designed races (restricted or race suckling).

Nutritional wisdom
A term used to describe the ability of animals to seek out feeds to satisfy specific nutrient deficiencies. Calves do not possess it.

Opportunity cost
The economic value of a particular feed if it were sold on the open market.

Oesophageal groove
A small channel in the rumen wall controlled by muscles, which allows liquid feeds to by-pass the rumen for digestion directly in the abomasum.

Pancreas
An organ that produces enzymes to assist with digestion of milk products.

Passive immunity
Resistance against diseases passed from cow to calf via the immunoglobulins (Ig) in colostrum. It is also called acquired immunity.

Pellets
Commercially produced and pelleted mixtures of feeds specially formulated for rearing calves or feeding specific types of livestock. They are generally based on cereal grains and other concentrates, but can also include agro-industrial by-products. They generally include specific mineral and vitamins.

pH
A measure of the level of acidity.

Probiotics
Additives, usually bacteria, to improve the natural process of digestion.

Pulpy kidney (enterotoxaemia)
One of the clostridial diseases.

Pyloric sphincter
The valve at the end of the abomasum, which controls the movement of feed into the duodenum.

Quality assurance (QA) program
A structured set of best management practices (BMPs).

Rumen
The major stomach in adult ruminants containing millions of microbes that breakdown feed particles prior to digestion by the animal. It is underdeveloped and non-functional in newborn calves.

Rumen degradable protein (RDP)
Dietary protein that is broken down into ammonia by rumen microbes.

Rotavirus
Microbes called viruses, which cause scouring in calves.

Salmonella
Bacteria causing severe scouring in calves. These can also be transmitted to humans.

Scratch factor
A term used to describe the usefulness of fibre to stimulate rumen development in young calves.

Seven-in-one
A vaccine used to protect against clostridia and leptospirosis bacteria.

Starter
A name given to the first type of concentrate fed to calves during milk rearing.

TLC (tender loving care)
This recognition of calves' health and general wellbeing can lead to successful calf rearing.

Transition milk
The milk from freshly calved cows (following colostrum) that milk factories will not collect for the first few days post-calving. It is usually (and incorrectly) referred to as colostrum.

Trochar
A device to puncture the rumen wall to relieve the effects of bloat in calves.

Undegradable dietary protein (UDP)
Dietary protein that escapes microbial digestion in the rumen and is broken down by the animal in the abomasum or duodenum.

Wastage rate
A measure of losses in replacement heifers between birth and second calving.

Whey
The by-product of cheese making.

White muscle disease
One major symptom of selenium deficiency.

Withholding period
The number of days following drug administration before milk or meat can be sold from treated animals.

Zoonoses
Calf diseases that can be passed onto humans.

Abbreviations

mm	millimetre
cm	centimetre
m	metre
mg	milligram
kg	kilogram
g	gram
mL	millilitre
L	litre
ppm	parts per million
MJ	megajoule
hr	hour
yr	year
lb	pound
ft	foot
hd	head
$	dollar
c	cent

appendix 4

Further reading

Textbooks and manuals

Dairy Quality Assurance Centre (1998), *Quality Replacement Heifers. Growing Your Own Profits*, Iowa, US.

Davis, C. and Drackley, J. (1998), *The Development, Nutrition and Management of the Young Calf*, Iowa State University Press, Ames, US.

Donohue, G., Stewart, J. and Hill, J. (1984), *Calf Rearing Systems*, Victorian Department of Agriculture, Technical Report No.96, Melbourne.

Drevjany, L. (1986), *Towards Success in Heavy Calf Production*, Ministry of Agriculture and Food, Ontario, Canada.

Hides, S. (1992), *Dairy Farming in the Macalister Irrigation District*, Second Edition, Macalister Research Farm Cooperative, Maffra, Victoria.

Meat Research Committee (1996), *Best Practice Dairy Beef. From Planning to Feedlot Delivery – A Practical Guide to Dairy Beef Production*, Meat Research Corporation, Sydney.

Mitchell, D. (1981), *Calf Housing Handbook*, Scottish Farm Buildings Investigation Unit, Aberdeen.

Moran, J. (1990), *Growing Calves for Pink Veal. A Guide to Rearing, Feeding and Managing Calves for Pink Veal Production in Victoria*, Victorian Department of Agriculture, Technical Report No.176, Melbourne.

Moran. J and McLean (2001), *Heifer Rearing. A Guide to Rearing Dairy Replacement Heifers in Australia*, Bolwarrah Press, Victoria.

National Research Council (1989), *Nutrient Requirements of Dairy Cattle*, Sixth Edition, National Academy Press, Washington, DC, US.

Natural Resources and Environment (1998), *Code of Accepted Farming Practice for the Welfare of Cattle*, Bureau of Animal Welfare, Agnote AG 0009, Melbourne.

New South Wales Department of Agriculture (1984), *Raising Dairy Calves*, NSW Department of Agriculture Agfact, A1.2.2, Sydney.

Roy, J. (1980), *The Calf*, Fourth Edition, Butterworths, Sydney.

Roy, J. (1990), *The Calf, Vol.1. Management of Health*, Fifth Edition, Butterworths, Sydney.

Schrag, L. (1982), *Healthy Calves, Healthy Cattle*, Verglag Ludwig Schrober, Auenstein, West Germany.

Standing Committee on Agriculture, Animal Health Committee (1992), *Australian Model Code of Practice for the Welfare of Animals. Cattle*, SCA Report Series No.39, CSIRO, Melbourne.

Tasmanian Department of Primary Industries, (1991), *Raising Dairy Replacements. A Manual for Dairy Farmers*, Department of Primary Industries, Hobart.

Thickett, B., Mitchell, D. and Hallows, B. (1988), *Calf Rearing*, Farming Press, Ipswich, England.

Webster, J. (1984), *Calf Husbandry, Health and Welfare*, Granada, Sydney.

Winter, K. and Lachance, B. (1983), *Managing and Feeding of Young Dairy Animals*, Publication 1432E, Canada Department of Agriculture, Ottawa, Canada.

Wishart, L. (1983), *The Dairy Calf in Queensland*, Queensland Department of Primary Industries, Brisbane.

Useful websites on calf rearing

www.animal_welfare.org.au: developed by Animal Welfare Centre, Melbourne.

www.australiancalfrearingresearchcentre.com (also www.acrrc.com): Australian Calf Rearing Research Centre.

www.babcock.cals.wisc.edu: Babcock Institute for International Dairy Research and Development, located at University of Wisconsin.

www.calfcountry.co.nz: Great Hage Company, New Zealand resellers of dairy products, with useful veterinary advice and an electronic discussion group.

www.calfnotes.com: American Protein Company, with many good calf technical articles including full text of 'Calving Ease' newsletter (monthly US discussion group specifically on calf rearing). Also has good hot links to US calf and heifer-related websites.

www.calfrearers.asn.au: Professional Calf Rearers Association of Australia, with good links to many overseas calf rearing organisations and technical articles.

www.dairyweb.com.au: developed by Dairy Research and Development Corporation.

www.grober.com: Grober Animal Nutrition, manufacturers of calf milk replacers in US, with links to technical articles on feeding calf milk replacers.

www.midlanz.com: Midland company, manufacturers of colostrum and blood testing kits.

www.pdhga.org: Professional Dairy Heifer Growers Association distributed to all PDGHA members and associated organisations. Membership is available by contacting PDHGA, 11 North Dunlap Ave., Savoy, IL 61874, US.

www.target10.com.au: Victoria-wide dairy extension program.

Electronic discussion groups

Some of the above websites include an electronic discussion group allowing anyone to send in a question to a calf specialist or veterinarian. There are other dairy-related groups that people can subscribe to:

Dairy-L is a US site to which people can subscribe. To join, send an email to listserv@umdd.umd.edu.

VicDairy-L is a free, Victorian-based, email discussion service. The contact is Frank Tyndall on (03) 5662 3502 or vicdairy-l@unimelb.edu.au.

Calf rearing newsletter

Calving Ease is a monthly newsletter produced by two US calf rearing specialists, Sam Lealey and Pam Sojda. It can be subscribed to by contacting Sam on sleadly@frontiernet.net or Pam on pams91@2ki.net. A limited number of back issues can be accessed at www.calfnotes.com.